M. Ya. Burlev
N.S. Nikolaev
S.A. Ryzhov

Scientific and practical bases of drying in the electric field

M. Ya. Burlev
N.S. Nikolaev
S.A. Ryzhov

Scientific and practical bases of drying in the electric field

Fundamentals, concepts, techniques, design, patents

LAP LAMBERT Academic Publishing

Impressum / Imprint

Bibliografische Information der Deutschen Nationalbibliothek: Die Deutsche Nationalbibliothek verzeichnet diese Publikation in der Deutschen Nationalbibliografie; detaillierte bibliografische Daten sind im Internet über http://dnb.d-nb.de abrufbar.

Alle in diesem Buch genannten Marken und Produktnamen unterliegen warenzeichen-, marken- oder patentrechtlichem Schutz bzw. sind Warenzeichen oder eingetragene Warenzeichen der jeweiligen Inhaber. Die Wiedergabe von Marken, Produktnamen, Gebrauchsnamen, Handelsnamen, Warenbezeichnungen u.s.w. in diesem Werk berechtigt auch ohne besondere Kennzeichnung nicht zu der Annahme, dass solche Namen im Sinne der Warenzeichen- und Markenschutzgesetzgebung als frei zu betrachten wären und daher von jedermann benutzt werden dürften.

Bibliographic information published by the Deutsche Nationalbibliothek: The Deutsche Nationalbibliothek lists this publication in the Deutsche Nationalbibliografie; detailed bibliographic data are available in the Internet at http://dnb.d-nb.de.

Any brand names and product names mentioned in this book are subject to trademark, brand or patent protection and are trademarks or registered trademarks of their respective holders. The use of brand names, product names, common names, trade names, product descriptions etc. even without a particular marking in this work is in no way to be construed to mean that such names may be regarded as unrestricted in respect of trademark and brand protection legislation and could thus be used by anyone.

Coverbild / Cover image: www.ingimage.com

Verlag / Publisher:
LAP LAMBERT Academic Publishing
ist ein Imprint der / is a trademark of
OmniScriptum GmbH & Co. KG
Heinrich-Böcking-Str. 6-8, 66121 Saarbrücken, Deutschland / Germany
Email: info@lap-publishing.com

Herstellung: siehe letzte Seite /
Printed at: see last page
ISBN: 978-3-659-67067-1

Zugl. / Approved by: Moscow, The Gorbatov's All Russian Meat Research Institute, 2014

CONTENT

INTRODUCTION

Intensification of technological processes based on the latest achievements of the science and the technology is the actual aspects of the food industry in the world. Technical progress, on the one hand, offers prospective ways and techniques to solve this problem, and on the other, makes it more difficult for her, as is accompanied by many negative phenomena.

Therefore, intensification of technological processes, including the convective drying of biological objects, at the present stage is the complex problem, in addition to product quality, addressing such issues as the energy efficiency, the environmental protection, the food safety, as well as with a number of technical and socio-economic issues.

Today, in addition to the technical progress in the solution of scientific and technological tasks significantly affects the socio-economic situation in the world, which is the new factor in determining requirements for industrial production and prospects for their development, including the food industry.

As one of the oldest areas of human activity, food and problems of processing of the milk in particular, in this context, do not lose their relevance. For example, to compete in the production of the milk and dairy products, as well as provide food security in this vital product necessary structural adaptation of all conditions of agricultural industrial processing of raw materials to the final output of dairy products on the market in industry countries.

Increased globalization trends will cause to the world scientific and technical developments, as well as the movement of the capital, often owned by large international companies will be deeper penetrate in the production process.

In the long term, this should lead to the acceleration of scientific and technological progress in the industry, the establishment of modern enterprises or the reconstruction fabrics and accelerate the development of new program technologies, etc.

Globalization is the objective competition and increases the concentration of the industry, that is, essentially contributes to the reduction of the number of enterprises. With regard to the impact of the globalization on development research, it contributes to the development of the work at the level of international programs and studies.

In addition, a quick update, the qualitative improvement of material and technical basis of the food industry, for example in Russia, by the use of the scientific and technological progress are key to reducing the cost of food production and to increase productivity, improve profitability and the capital manufacturing. Intensification of the drying process and determine the most optimal regimes are the issues that need to solve to create the highly efficient industrial food equipment.

Dry biological products are used in many industries, fish, flour-grinding industry, not only food, but also of other equally important production plants of different profile in Russia, for example in the medicine, the pharmaceutics, the microbiology, etc. The timber industry also used the technological process of wood drying in the world.

However in this research, the emphasis in studies make on the intensification of the process of drying of the skimmed milk and improving technology and the technology of its production.

Development of spray drying of biological objects is based on the fundamental works of Soviet and Russian scientists: Rebinder P.A., Lykov A.V., Lykov M.V., Ginsburg A.S., Lurie M.Yu., Cook G.A. and foreign scientists: Masters K. (United Kingdom), Buma J. (Netherlands) and Taneva S. (Japan) and others [24, 28, 55, 57, 58, 73 and 86].

A great contribution to the development of the technique and the technology of the drying process of foods and development of heat-mass exchange processes have made modern scientists of Russia: Lipatov N.N., Kharitonov V.D., Ivashov V.I., Ilyukhin V.V. Rudobashta S.P. and others [32, 45, 49, 53, 79, 84 and 85].

Analysis of different energy impacts on biological objects showed that the electrical technology should continue because of their potential for intensification of the drying process. The special attention deserves the electron-ion technology, namely the effect process impact of the electric field directly on the food product, or via an intermediate medium, for example - the flowing air.

Given the great need for dry foods and the presence of the high-energy intensity of the drying process, questions intensified, the search for new technical and technological solutions, which increase the efficiency of the process and research in these areas, are very relevant.

Chapter 1   *ANALISIS OF WAYS TO ENHACE THE EFFECTIVENESS OF THE DRYING PROCESS OF BIOLOGICAL OBJECTS*

1.1.   *ANALIS OF THE BASIC WAYS OF THE DRYING BIOLOGICAL OBJECTS*

The evaporation of moisture from the drying of biological objects there is concentration gradient of the moisture, which is the driving force behind the movement of internal layers to the surface of the object. The migration process and moisture removal violates its connection in the biological materials and requires application considerable energy [27].

There are various industrial technological methods of a supply of various type of the energy for removal of the moisture from the dried-up biological objects, for example, the contact, the convective, the thermo-radiation, the ultrasonic (acoustic), the currents of high frequency in the electromagnetic field, the ultrahigh frequency (microwave oven and converters), etc. [1, 4, 7 and 9].

The principle of the contact supply of heat is used in drying of biological substances, for example, of the skimmed milk on drum or tape dryers at rather low power consumption.

Most common in the food industry are spray dryers, especially in dairy industries, which in this scientific work are the main object of research [78, 84]. One of the main directions of improving modern spray dryers is the development of methods intensification of the drying process and modernization of the structures, which provide high-quality food and dairy products.

Development of this direction is follow:

–   use of more progressive methods of processing of the biological products at various stages of technological process:

–   introduction to the hardware scheme of new elements of the equipment;

–   increase of extent of automation of process.

Separate modern development is direct not only on improvement of quality of biological objects, but also on creation of compact installations [53, 56 and 57].

Many biological substances and materials need to dry at the lowered temperature, because owing to the thermo-labile the increase in temperature often conducts to decrease in qualitative characteristics and properties. Under the vacuum is carrying out drying of thermo-labile substances [56].

The vacuum drying of biological objects in the frozen state with the pressure of the steam-gas mix less than 4,58 mm of mercury, often call the sublimation drying. In the course of this drying the frozen moisture, which is in biological object passes into the vaporous state, passing the phase of the liquid, which is the "sublimates".

It also call molecular, besides, the term "lyophilic" drying which is used generally for preservation of the main biological qualities and is applied in the medicine and in the pharmaceutical branch, for example, by production of the medication and other substances. The sublimation drying is applied to receiving the quality foodstuff possessing the increased biological value [31].

Duration, power consumption of the process and complexity of the sublimation equipment is to the main shortcomings of the sublimation drying in industry.

One of the most perspective ways increase efficiency of the drying is use acoustic and in particular ultrasonic fluctuations. At this way of the energy, surrounding of the biomaterial is exposing to vibration impact of ultrasonic fluctuations [14]. The accelerated removal of the moisture at the ultrasonic drying is provided with decrease in the diffusive resistance in the volume and from the surface of the material. It's reached due to action of powerful turbulent gas streams on material surfaces, and reduction the thickness of the surface layer.

As a result, the application of the technology of the ultrasonic drying is possible to reduce process temperature to the values ensuring safety of biologically active agents, quality of initial biological objects, to increase the drying process speed, to reduce losses of the product.

The equipment for the ultrasonic drying easily adapts for traditional drying installations with the vibration layer, the tunnel, the drum and another, significantly increasing their productivity and economic efficiency.

7

However, the acoustic drying has serious shortcomings:
- the low efficiency, which is not exceeding 20%;
- fast wear of mechanical knots;
- impossibility of work at high frequencies (more than 20 kHz) and, as a result, need the protection of the service personnel [69];
- practically 80% of the acoustic energy are absorbed by additional devices, and does not participate in the drying process.

All listed shortcomings reduce efficiency of the acoustic influence and do not provide the acceptable speed of the ultrasonic drying.

For removal of the moisture from thick-layer biological substances when it is necessary to adjust temperature characteristics and the gradient of the moisture on the surface and in substance, apply drying with use of the currents of the high frequency. In such a way, it is possible to dry, in particular, the plastics and other biological materials possessing dielectric properties [54].

Under the influence of the electrical field of ultrahigh frequency parameters, positive ions and electrons in the biological material (the electrolyte containing structurally certain quantity, for example, of the gradient solution of salts) change the direction. At this process of the drying energy in the form of the heat use for intensification of the dehydration.

If changing parameters of potential electrical field, namely its intensity, it is possible to influence temperature fields in the material and by that to control the high-speed modes in the process of dehydration.

Also used of the thermo-radiation dryer, due to the heat from infrared rays in industry.

Drying of damp biological substances, at which warmly, necessary for evaporation of the liquid and heating of the material, which is transferee in the basic by the radiant energy, usually call the radiation or drying infrared rays.

Sometimes distinguish drying by sub infrared rays that is generators of the radiation are special lamps, which, besides infrared rays, radiate visible rays.

In this case the drying by infrared rays will happen when using the generators of radiation, heated lower than temperature of the luminescence. Usually the drying by thermal or thermal radiation is follow by drying heated gas. Therefore, radiation convective drying in most cases takes place [19].

In the specified way, it is possible to bring the specific streams of heat falling on one square meter of its surface, in 10 times exceeding to the biological material the corresponding streams at the convective and the contact drying.

Therefore, when infrared drying applied, increases the evaporation rate the moisture of the biological object.

So radiant heating at drying is effective primarily for the drying of biological objects in the thin layer. However, the drying by infrared rays (lamp dryers) has a limited scope and use where the main heat power indicators (expenses of the heat and the electrical power) have no crucial importance in comparison with duration the process of the drying and quality of a ready-made product [53].

The main disadvantage of thermo-radiation dryers is that they have a high energy consumption is 1.5-2.5 kW•h/kg of evaporated moisture, which limits their application. Practically all known methods of the drying, differ in high power expenses owing to what search of various ways of an intensification of processes of the drying. It is very actual task and demands special additional scientific researches [79].

Especially it belongs to the least studied sphere of an intensification of process of drying with use of weak electric impulses in laboratory experiences and industry.

1.2. THE USE OF THE ELECTRON-ION TECHNOLOGY
IN INDUSTRIAL PRODUCTION
PROCESSES

In recent years, along with become already traditional technological spheres of the electricity in the industry was created the new direction called the electron-ion technology. It is basing on use of forces of interaction of the electrical field, electrical potential and electrical charges for the organization of the ordered movement of particles of substances and materials [10].

The electron-ion technology is in "combination" of several directions of physics and production technological processes: electro-separations, the convective and the contact drying, electrostatics, the refrigerating equipment, electrodynamics and equipment of high voltage, mechanics and gas dynamics. Though the level of development each of these areas rather high, however sharing of these directions is still insufficiently studied in sciences and industry [46].

Laws of behavior of charges in the electrical field define the movement of particles of substances and in this regard, it is similar to the movement of electrons and ions in electronic devices. Along with this similarity there is also an essential difference which is that processes of the electron-ion technology represent the processes of electronics, which transferred electron functions.

It apply to the industrial devices and, respectively, from vacuum in the atmosphere. Carriers of charges in this case are not only electrons and gas ions, but also charged particles of substance. All these circumstances complicate character of the occurring physical phenomena. In the industry and the laboratory the electron-ion technology uses the electrical field in the form the action of mechanical forces on electrically charged substances and materials.

The electrical field is creating by connection of electrodes to the industry generator of the high voltage [27, 76].

The founder of Russian the electron-ion technology in the industry was Professor A.L. Chizhevsky, who made the big contribution to the theory and practice of such branches, as the curative medicine, the pharmacology and the agriculture [22]. In the agriculture by numerous researches, it was prove that possibilities of the electron-ion technology find the application in the plant growing, the animal husbandry, the vegetable growing, the gardening and other branches [5].

In 1970 years at the Moscow institute of Steel and Alloys were make technical and economic researches of application of the electron-ion technology at the enterprises. However, this development concerned electrical separation of raw materials and substances [30].

In the food industry, in particular in the field the processing of meat products the electron-ion technology (high-voltage ionization) for electro smoking of meat and electrostatic purification of gases [77] is used. In the dairy industry, the electron-ion technology was applied in the drying equipment for purification by the air (the aero-suspension) of the smallest particles of the powdered milk. Conditions of effective work of all drying installations in many respects depend on the correct functioning of filters. Attempts to use electrostatic filters were make, however they were not widely adopted in the technology the drying of dairy products [53].

Having analyzed the researches devoted to process of the drying with application of special methods it is possible to note that use of certain energy influence or on the dried-up biological material, or on the thermal agent is their cornerstone.

Practically in all cases, there are both positive and negative moments, essential to this process. Researchers solved the problem of saving of qualitative characteristics of the dried-up biological objects and it is the positive moment.

Negative factors consist in considerable use of the energy component (creation of stability of an ultrasonic wave, control the current of high and ultrahigh frequency, complexity of the thermal radiation, energy consumption of sublimation) which is required for the drying of biological objects. It is rise in price of constructive elements of the drying equipment in industrial conditions.

Scientific publications on research of influence of electrical technologies or various energy influences showed that studying in the field needs to be deepened and concretized taking into account both positive, and negative characteristics. Besides, studying of patent materials allowed considering ways and the equipment used when drying in the technological process and special attention to give to the electron-ion technology also, namely process of electrical impact directly on biological objects and other materials [65, 66].

Analyzing results, of information researches, can possible to do conclusions:

– one of the perspective directions of the intensification of drying process is application of the electrical field;

– works on research the new directions in the field the drying of biological objects on the basis of use of the electron-ion technology wasn't carried out.

Combination directions, expediently only based on application of system methods of the researches, which providing modeling of "weak" electrical impulses in the laboratory and then on the industrial production.

The perspective of studying of process of drying process with use of "weak" electrical impulses is actually, for example, for the drying of dairy products, in particular, of skimmed milk and other biological objects and substances. Current was measure in researches on the drying of biological objects in milli - Amperes therefore electrical impulses are call "weak".

Chapter 2   *THEORETICAL AND EXPERIMENTAL BASES AND*
REGULARITIES OF THE ELECTRICAL FIELD
ON PARAMETERS OF THE DRYING
PROCESSES

2.1. THE "CORONA" DISCHARGE AND ITS INFLUENCE ON PROPERTIES OF BIOLOGICAL OBJECTS AND THE ENVIRONMENT

Consideration of theoretical bases and regularities of influence of weak electrical impulses are necessary for studying and research of possibility of the electron-ion technology in relation to processing of agricultural biological objects (for example: milk and dairy products).

Before applying process the dehydration of biological objects with use of weak electrical impulses, it was necessary to study theoretically and to investigate almost full integrate component the action of the corona discharge. It was use as the first stage of studying the influence of the electrical field on the drying process.

In the technical and scientific literature is in detail described possibilities the application of the phenomenon of the "corona" discharge for increase in positive aspect of various parameters in such technological processes as, the dehydration of biological objects, the sublimation, cleaning of rooms by method of the antiseptic, the filtering, the melting and etc. [6, 23 and 33].

Unfortunately, in this technical and scientific literature there is no the detailed description concerning its main characteristics, chemical and physical aspects of the influence this phenomenon on food products, in particular, milk and dairy products. Unclear, what negative radiations of the "corona" discharge harmful influence a human body.

This process is in an area of the investigation. It is necessary to resolve issues of the ecological protection in the atmosphere and the environment.

Scientist A.L. Chizhevsky in the scientific publications and books [22] warned more than 80 years ago about illiterate use of various sources, so-called the "ionizers", which in his opinion, cannot productively induce necessary the gas ozone for treatment of the human body and, thus, it is possible to do considerable harm and to the person, and the atmosphere surrounding us.

Based on studying of the offered options use of the "corona" discharge phenomenon it is possible to draw some necessary conclusions, which will be necessary for further idea of consequences of its application:

– changes qualitative characteristics components of air, namely, there is the education not only ozone, but also with vapors of the water are produced nitrogen oxides harmful to the human body and the produce production;

– influences of powerful electrical fields in the course the induction of the electrical impulses are still poorly investigated. Namely - collateral negative consequences, as in the direction of powerful electromagnetic characteristics, and possible x-ray radiations. It occurs owing to the ionization components of the air, which is characterize by endothermic process the formation ions of neutral atoms or molecules;

– the static electricity in the course the induction of the electrical field and frequencies of electrical impulses is not studied.

In one cubic meter of the atmospheric air, there is a significant amount of various micro components, each of which can change considerably the structural component at the molecular level and it is necessary to consider it. It is necessary to investigate process of external influence the source of the "corona" discharge.

The main researches have to be direct on the extent influence of the ozone. In the course the action of the "corona" discharge is formed the oxide of the nitrogen, which transformed to the aggressive nitric acid. It is know that the nitric acid can destroy the structural component of physical and chemical characteristics of various minerals and substances.

The nitric acid can negatively influence environment, the industrial service personnel, neutralize sensitive touch technical parts, the sensor control - measuring devices, and industrial details of the processing equipment also. Besides, it is worth to remember about the ecology of the atmosphere or the air space. As atoms, molecules and negative radiations of microscopic substances, that is impurity, have considerable influence on various phenomena, which surround us [3].

Very early researches the method on receiving the oxide of the nitrogen from the air, as the result use of the "electrical" discharge were conducted American and English by scientists Cavendish and Priestley in 1781 – 1784 years [20].

The French researcher Lefebvre receives the first patent for the way receiving oxides of the nitrogen by means of the electrical discharges in the air and their transformations into nitrates and nitrites in 1903 year with use in industry [25]. Until 1925 year at the Baden plant, the firm BASF (Germany) the industrial method receiving from the air oxides of the nitrogen and the nitric acid by means of the electrical spark discharge was applied [51].

For example, there is the interesting publication of the German scientific and technical source, which informs that, according the State Federal laws of the Federal Republic of Germany use of gases in the form of the ozone for processing different of products in particular of meat products and semi-finished sub - products from the meat are really forbidden [52].

The essence in that properties of meat products and forcemeats due to influence of the ionized air with formation of a bright red shade on its surface considerably change and appears the false perception buyers, that these meat products fresh absolutely.

In other scientific publication the condition of the human body in which there is the oxygen starvation because one of forms of the hemoglobin can be changed ("oxidized"), because of the influence of the oxide of the nitrogen (one of forms of the nitric acid) which negatively influences on the nervous system is described [29].

Components of the nitric acid, in particular, the nitrogen oxide as the same publication describes, can be neutralize due to the change of the temperature. However, when applying the negative temperature is its preservation. It speaks about inadmissibility of the use of the process of "corona" discharge in the industrial conditions of the refrigerating and freezers [83].

The Russian researcher wrote in their monographs to use the ozone for anti-septic warehousing, workers have to have the protective masks or not to be in limits the influence of ozone until 7 hours [90]. Confirmed, that the quantity of viruses and microbes due to impact of air by the ozone is decreasing. Therefore, the period the storage of products with the subsequent its realization considerably increases.

However, there are also other considerable advantages of use process of the ozonation in industrial conditions, for example: in the medicine, the agriculture, in the pharmaceutical industry, the applied biotechnology, refrigeration enterprisers and other branches [23, 60]:

- the ozone about 300 times faster than other disinfecting substances destroys all microscopic organisms, known in the nature;
- the ozone is the good disinfector and can apply in any technological production at the enterprisers;

- the ozone can show the high efficiency in any technological production in the short interval on time;
- the ozone is good "deodorant" and also it can be applied in the cosmetic industry and in production;
- the ozone does not influence on "pH" H_2O and does not synthesize components, from water molecules, which are negative for environment;
- the ozone does not form toxic by-products and is transformed at the molecular level to O_2 ;
- fast back in the air atmosphere;
- the ozone does not form toxic by-products;
- the ozone is develops on the place, without demanding storage and transportation in industrial conditions [88].

2.2. STUDY ON PROPERTIES OF THE ELECTRICAL "CORONA" DISCHARGE IN THE ATMOSPHERE

Properties of the electrical "corona" discharge in air influence processes of the electron-ion technology. With the one side, the electrical conductivity of air limits the maximum the voltage of the electrical field and by that force to operate on particles.

With another side, the "corona" electrical discharge in the air allows carrying out effective charging of particles and therefore it is essential to expand possibilities of the process [6, 8].

In the air in the result of various external atmospheric actions and the influence (the ultra-violet radiation of the sun, atmospheric phenomena space rays, etc.), always there is a quantity of the ions and electrons reporting to the air a certain conductivity.

For example, in one cubic centimeter of the atmospheric air some ten thousands of couples of ions, which through certain time can recombine with each other, are create every second and again turn (to recombine) into neutral molecules.

If to put the constant voltages to the air interval between flat electrodes are the volume of the voltage can be significantly change, ions will start moving along basic power lines of the electrical field, forcing electrical current in an external contour [87].

In the process of increase, the voltage of the electrical current increases because the most part of ions manages to reach electrodes, without recombining in the air. The regime of saturation finish when ions in the air practically do not recombine.

At further increase in the high voltage electrical current starts over again to increase that testifies to the begun ionization of the air under the influence of the electrical field. At some value of the voltage, there is a sharp increase of the current, which indicates sudden high-quality change of the condition of the air atmosphere.

This high voltage is call as the "impulse breakdown" voltage of the air interval, which is at achievement of this voltage atmospheric air loses properties of dielectric and turns into the conductor.

The electrical "corona" discharge, or the "corona", is a peculiar form of the electrical discharge, characteristic for strongly fields when ionization of air happens only in small area near an electrode.

In the interval between the electrodes do not occur through the conductive channel. Therefore, education of the "corona" discharge does not mean total loss of air gap insulating properties.

Therefore, the focus has been on creating the most appropriate construction of the contour electrode to continue further research in the field of electric field.

2.3. THE DEVELOPMENT AND RESEARCH OF THE OPTIMUM CONSTRUTION OF THE CONTOUR ELECTRODE "ANTENNA-EMITTER" AND INDUSTRIAL DEVELOPMENT THE IMPULSES GENERATOR OF THE HIGH VOLTAGES

In accordance with new ideas about the physics the influence of the electrical field on the biological object developed and continually changed the construction "Ionizer", which essentially became the "antenna-emitter" in the form of the contour electrode. Typical electrodes – "antennas-emitter" for creation of the form and transfer of the electrical discharge are thin electrode threads (the string), the sharp needles or close-meshed screen grids (Fig. 1).

At the application of the certain voltage to electrodes, the volume of voltage of the field for emergence of "corona" takes place only in the area where there are electrode threads. The quantity and interval between of electrode threads define proceeding from experiences at which there is the electrical discharge

The voltage of the field is especially great near the "antenna-emitter" electrode which contour is execute in the form of electrode threads with the diameter less than two millimeters, and at the application of other electrodes in the form of the sharp needles or the screen grid – sharply decreases.

The electrical "corona" discharge is localizing on this electrode, which calls the "corona-electrode". Emergence of the "corona" requires a certain degree of heterogeneity of the electrical field. With big diameters of electrode threads in the contour electrode "corona" is the unstable phenomenon and will be only a short-term stage before emergence of the "impulse breakdown".

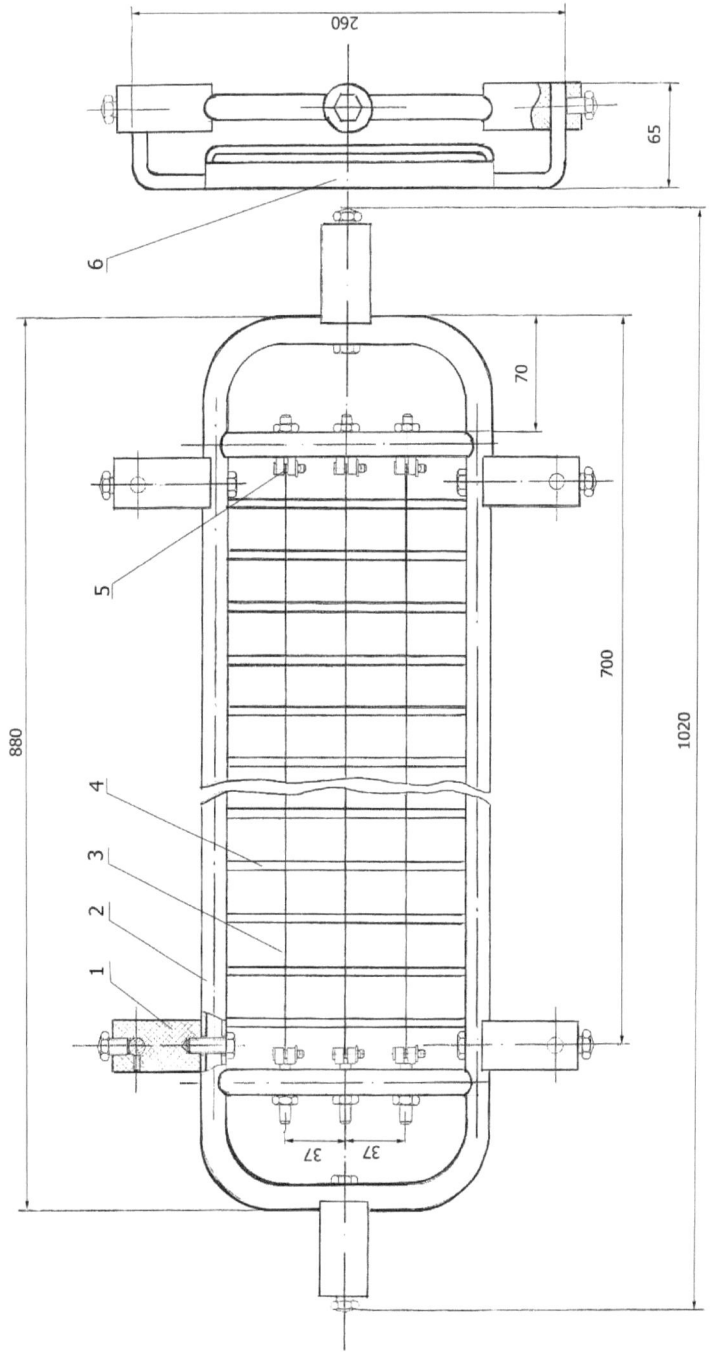

Fig. 1 The contour electrode "antenna-emitter":
1 – the insulator; 2 – the frame; 3 – electrode threads; 4 – the electrode contour of the opposite sign;
5 – the mechanism tension of electrode threads; 6 – the isolating frame of the electrode contour

There is the rule: the heterogeneity of the field is less, the divergence between the voltage emergence of the "corona" and the voltage of the "impulse breakdown" is more. Thus, other things being equal the probability the emergence of the "corona" is more, than thinner threads or more sharply the end of needles of the electrode.

The "corona" discharge there are corresponds the increasing part of the Volt-Ampere characteristic, therefore that, there was the "impulse breakdown" between electrodes, it is necessary to increase the voltage.

The ionization and processes of the recombination and transition of the excited molecules to a normal state accompanying it promote formation of quanta of the light due to what a peculiar shine round the electrode or the contour electrode "antenna-emitter" is created, from here and there was the name – "corona".

The carriers of the electrical charge, which are form in the ionization zone under the influence forces of the field, move in the field that is there is the electrical wind of the "corona" discharge. Ability of the "corona" discharge is the source of large number carriers of charges of one sign and transfers it through the air to the charges surface.

This gives the chance to apply the "corona" discharge, for example, in electrical separation for the charging of the separated particles.

At the "corona" discharge with emergence of the electrical current is the electrical field between electrodes becomes dynamic and for this case it is impossible to calculate of the voltage of the "corona" discharge, by the formulas suitable for the calculation of the high voltage in the electrostatic field [47, 48].

The voltage of the "corona" of the electric current represents in the kind of the phenomenon interrupting, which was measured by oscilloscope, which observed in the form of impulses, lasting about 10^{-8} second, about 2×10^9 couple's electrons and positive ions.

That is in process of increase voltage the frequency of impulses can increases from units to several hundred kilohertz.

If to change of polarity on the contour electrode "antenna-emitter", certain phenomena are observed. For example, impulses the current of the "corona" discharge from the electrode at negative polarity arise at the time the achievement of the initial voltage of the "corona" discharge and continue to exist at the further increase of the voltage.

Moreover, impulses from the electrical current of the "corona" discharge from the electrode at the positive polarity arise, also at the initial voltage and, exist only in very narrow diapason of voltage exceeding the initial voltage. In these conditions regularly, repeating flashes of the "corona" discharge are observed. At the subsequent increase of the voltage are impulses of the current merge in continuous electrical current of the "corona" discharge.

Character of the luminescence round the "corona" electrode is connecting with features the development of the charge at various polarities. In case of the negative "corona" discharge on the surface of the conductor or the electrode (threads in the form of strings) there are separate, all the time moving shining points. In case of the positive, the "corona" discharges the luminescence on the surface of the conductor or electrode (threads in the form of strings) more evenly. In the direction forces of the electric field, ions and electrons on length of free run are moving evenly.

These are accelerated and accumulate the kinetic energy, however in the process of collisions this kinetic energy is partially or completely transfer to other particles. That is the movement of ions or electrons in the air similar to the movement with friction. In very weak electrical fields in the process of mutual collisions the accelerated ions and electrons lose only part of the saved-up the kinetic energy.

But at some critical voltage of the electrical field, disseminated in collisions kinetic energy increases so that the average speed of the movement of ions and electrons is stabilized [59, 81].

Constant speed of ions and electrons "V" is characteristic, in particular, for the "corona" discharge at which it is proportional of the voltage electrical field:

$$V = RE, \qquad (1)$$

Where; R – coefficient of proportionality or as it is called still, coefficient of mobility of ions, cm^2/(Volt·sec).

E – coefficient of equal to average speed of the movement of ions and electrons, Volt/cm.

The active mobility of ions of both signs at the time of their emergence in the air with a pressure 10^5 Pascal. and temperature 293^0 by Calvin is identical and equal about 2,5 cm^2/(Volt sec), however already through 10^{-2} second, (after 10^7 collisions) the mobility decreases and becomes for positive and negative ions respectively:

- $R_+ = 1,6$ cm^2/(Volt sec);
- $R_- = 2,2$ cm^2/(Volt sec)

Decrease the active mobility of the ion happens because of its sticking to the molecule of water and increase thereof the mass of the formed complex – the heavy ion. Vapors of the water have low the mobility 0,6 cm^2/(Volt sec) at 293^0 by Calvin, and complicates development of ionization processes. In the presence, vapors of the water the active mobility of negative ions are decrease.

It is known that one of the most important characteristics and properties of the "corona" discharge is dependence the current of the "corona" discharge from the volume voltage, which is the Volt-Ampere characteristic (Fig. 2).

This dependence shows the potential of "corona" electrodes of different construction (electrode threads, the needle frame, the close-meshed screen grid) is more effective, which depend from various substances and materials (the copper, the aluminum, the metal, the silver, the platinum, the wolfram, etc.).

Thus, it is possible to determine the power consumed by current of the corona discharge. Electrical current of the corona discharge is more, than nearest working tension to "impulse breakdown" therefore for increase of charging of particles. It is necessary to support the tension on electrodes at the highest level. In case of "impulse breakdown" of the "corona" interval, current of the "corona" increases in hundreds of times.

For example, if electrodes are metal needles with a diameter 0,1 mm and the surface of the plane, the corona discharge arises at a potential difference more than 1000 V, and is followed by current 0,05 A, and at "breakdown" current reaches several hundred Amperes.

To operate such process very difficult, therefore, thin threads in the contour electrode were chose. Current of the "corona" discharge depends not only on a type of the electrode (an electrode thread, needle or a screen grid), but also on a relative positioning of electrodes, their polarity and various materials of the contour electrode, by name of "antenna-emitter" (copper, aluminum, iron, silver, platinum, etc.).

For the construction of the contour electrode "antenna-emitter" is selected optimum number of electrode threads (strings), which show on Fig. 1 and Fig. 3. If increase number of electrode threads are the electrical field to be non-uniform and the "corona" disappears.

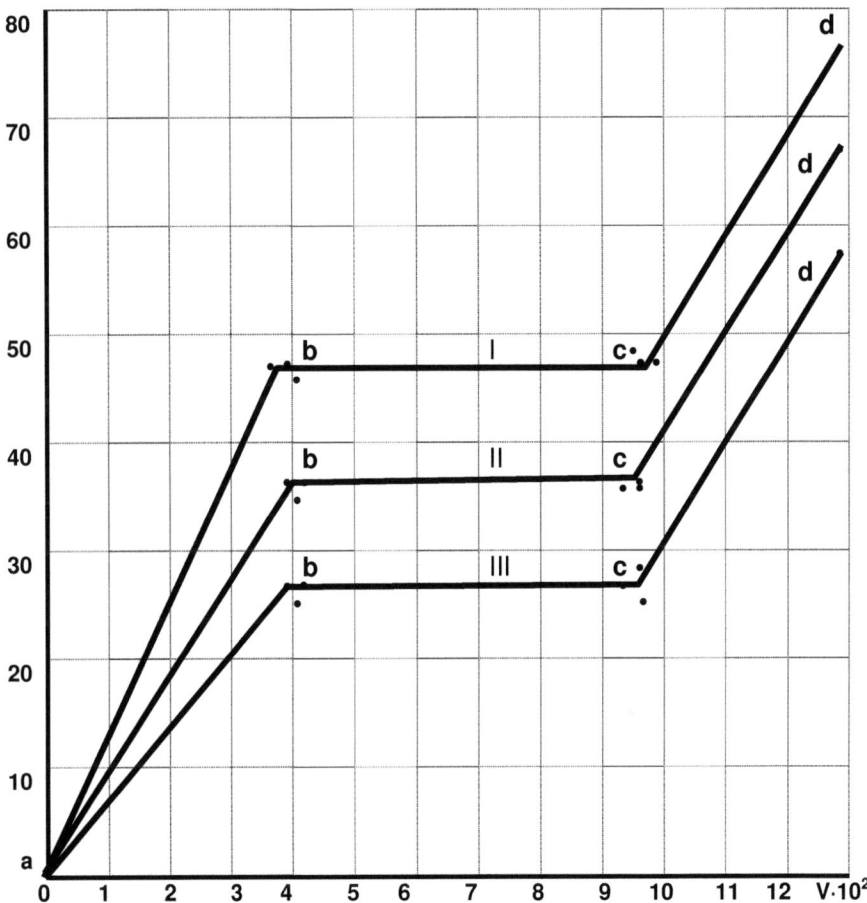

I·x 10⁴, Ampere

Fig. 2 Volt-Ampere characteristic of the electrical "corona" discharge:

I – the electrode in the form of the sharp steel needles;

II – the electrode in the form of a copper screen grid;

III – the electrode in the form of thin electrode threads, material silver.

"a – b" – area of increase of current;

"b – c" – saturation current area;

"c – d" – ionization area.

It is very important factor for further research. The Volt-Ampere characteristic of the electrical "corona" discharge depends also from the temperature of the environment and the humidity of the air. At increase the humidity of the air or decrease of the temperature are current of the electrical field decreases and on the contrary

For example, it is consider when carrying out process of the electrical separation, as fluctuations of the humidity lead to essential change of technological indicators [2]. It is know that amount of electricity of the transferring environment proportionally to the square of electrodes and the coefficient of proportionality depending on dielectric permeability of the environment filling space between the contour electrode of the "antenna-emitter" and the electrode of other polarity [76].

As the air environment possesses rather big resistance, electrical energy accumulates on the contour electrode of "antenna-emitter" the construction. This energy in researching is characterized by very weak fields and respectively the currents measured in most cases in milli and even in micro-Amperes, and the voltage in Volts and milli-Volts. However, if it necessary, these parameters can be increased [34].

In the laboratory of the Moscow State University Applied Biotechnology more than twenty-five years researches on use and industrial application of the electronic-ion technology for an intensification of technological processes when processing biological objects were conducts. Professor V.V. Ilyukhin headed the group of researchers. Thus, negative influence (impact) of the air in the process using of the electrical "corona" discharge was consider [16, 46].

The sample with the blood on the technological production of the bactericidal medicines and preparations at the Scientific Research Institute of Epidemiology and Microbiology in Moscow was investigated. The ionization air processed the blood of the animal.

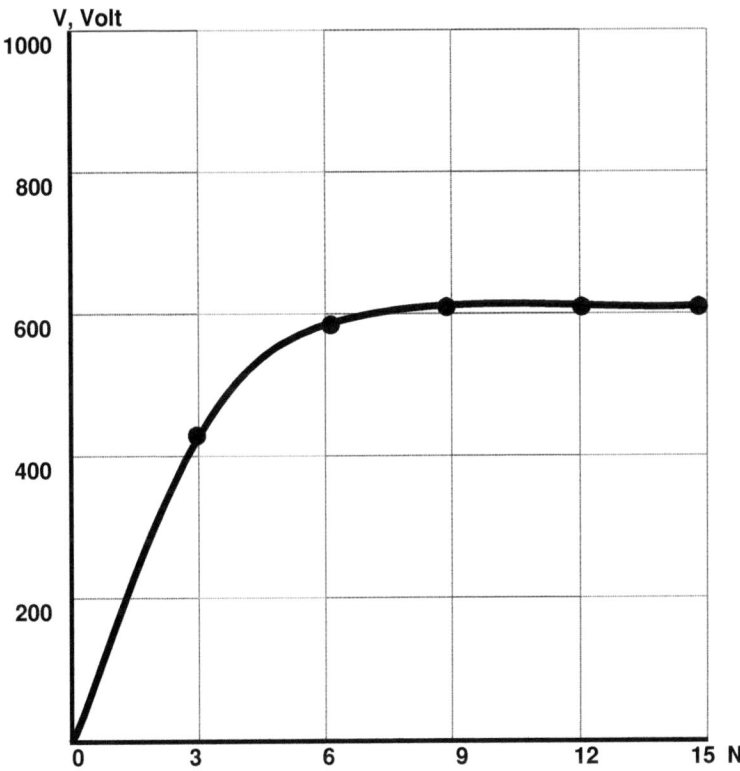

Fig. 3   Dependence the voltage of the electrical field from quantity
of electrode threads on the contour electrode
"antenna-emitter"

The current - 0,028 Ampere,
The frequency of impulses of F = 200 Hertz
The duty cycle of impulses S = 3

There was a destruction of erythrocytes, which is a component of the liquid composition of the blood, but with changed of the color of the blood, close to discoloration. This process happened during ten – fifteen minutes. It means that, the reaction of the interaction of the blood with the air processed by the ionization air happens very quickly. It is negative factors of the electrical "corona" discharge.

Professor G.K. Berman, from the Scientific Research Institute of the Rubber and the Latex in Moscow conducted researching of the new way or the method the process the drying of the latex. He found out that, the ionization air processed by the electrical "corona" discharge, within 7 – 8 minutes destroys the surface of samples from the latex. Thus, deep micro-cracks are form on all volume of the latex production.

The oxide of the nitrogen was the reason of the destruction of polymer substances and materials [67].

In general, based on theoretical and own experiments the influence of the "corona" discharge on biological objects was formulated the scientific hypothesis, which capable to explain the mechanism of the intensification the process of the drying at influence of the electrical field.

The idea of the hypothesis consists in the following: the intensification the process of the drying of biological objects and substances with using the weak electrical impulses it is possible to explain with emergence of such micro-electro-kinetic processes, as the micro-electro-osmosis, the micro-electrophoresis and the micro-electrolysis [21, 75 and 89]. However, use of the electrical field in the form of the "corona" discharge is inadmissible from the ecological point of view and from the safety measures position. Besides, contact of the ionization air processed by the "corona" discharge with the food product is undesirable in connection, with the oxidation of the fat, due to influence of oxides of the nitrogen, the ozone, the nitric and other synthesizable acids.

Experimentally established that, the impulse oscillations (modulation) of the electrical current, induced with the drying of materials or substances materials, which can to initiate on their surface the totality of whole series of processes: for example - micro-electro-kinetic phenomena (the micro-electro-osmosis, the micro-electrophoresis), the micro-electrolysis and the really stochastic resonance of its own impulses, generated by evaporating of the moisture and induced by means the electrostatic induction.

Therefore the new ecological way or the method and the device inducing electrical charges for the intensification of the processes of heat- and the mass transfer processes was developed is the technology the using the influence of weak electric impulses [15].

Basic versions of schemes of the organization of application of electrical technologies are present on Fig. 4. Besides, the device, which enters in the industry technology the process of the drying of biological objects or substances, is developed and improved.

This industrial device is the generator of different impulses of the high voltage [63]. Performed at this stage of the studying allow the elaborate impulse high voltage generator, which served as the basis for further experiments.

The impulses generator provides the impulses amplitude, the frequency and the duty cycle in the proper range of characteristics. The high voltage impulses generator is the construction to produce the high voltage up to +1000 Volts in two channels, with the overall scheme of the modulation.

The modulation is switched (on/off-button "the modulation") output the high voltage is impulse sequence with the frequency from 50 to 1000 Hertz (adjustment knob "frequency modulation") and the outputs are the impulses modulation.

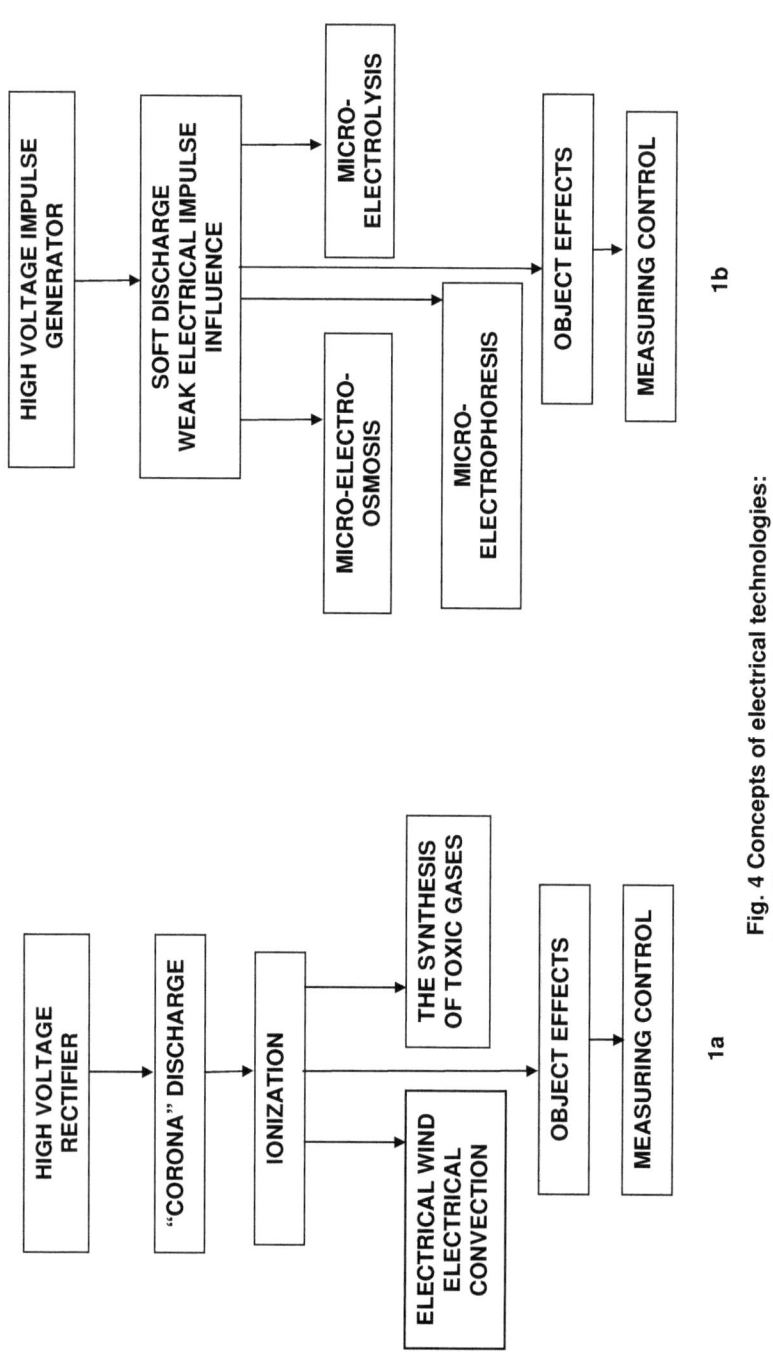

Fig. 4 Concepts of electrical technologies:
1a – The traditional way; 1b – The new way.

2.4. ANALYSIS THE OPERATION OF THE SYSTEM: GENERATOR– CONTOUR ELECTRODE "ANTENNA-EMITTER" –BIOLOGICAL OBJECTS – "ANTENNA-RECEIVER" AND THE PROTECTION OF ELECTRICAL MEASURING DEVICES AND EQUIPMENTS FROM THE ELECTRICAL RADIATION

In this chapter provides analysis of the system operation: high voltage impulse generator – contour electrode "antenna-emitter" – biological objects (condensed skimmed milk) – "antenna-receiver". The main subject of study food products (biological objects), as in subsequent experimental scientific works, selected the condensed skimmed milk.

In addition, in the form of the hypothesis is given for further explanation of the intensification of the process of the drying, using weak electrical impulses influences on biological objects of the drying, which is condensed skimmed milk.

The study finds that the underlying value for the drying process of the condensed skimmed milk in the scientist laboratory has induced electrical component in the form of the induction. When the intermediate distance between the contour electrode the "antenna-emitter and "antenna-receiver" is from 0 to 1 meter.

In industrial environments, the main role is assigned to "free substations" are transforming the electrical charges, the so-called "the electrical conduction current". The approximately distance between the contour electrode "antenna-emitter" (the ionizer) and "antenna-receiver", where as biological objects are the condensed skimmed milk, exceeds one meter or more. Under the electrical condition "free substations", means the air particles (dust) with the positive or negative charges, etc.

In the research laboratory of Moscow State University Applied Biotechnology has developed the number of high voltage impulses generators. The need for such development occurred as the physical model of the drying process by using the weak electrical impulses influence. First, apply the "corona" discharge generator by type "Chandelier of Chizhevsky". Everyone thought that the cause of the intensification of the process of the drying is the ionization of the air and the transfer of "ions", that is, the impact of the "electrical wind", the so-called "electrical convection" [22].

However, analysis the physics of the process, that led to the assumption, that the physical model is incorrect. In our research [64] was put forward the hypothesis that ".... and one of the main reasons for the increase of the drying process is "white noise" – the certain the frequency spectrum of electrical impulses, generated by the electrical "corona" discharge, that occurs in the certain frequency range of impulses and impulse duty cycle [35].

In addition, "is ... under the influence of the stochastic resonance (the stochastic resonance is a convergence of natural frequencies of impulses of the object (for example, the condensed skimmed milk) with chaotic dynamics induced by impulse frequency from the external source) speeds up the drying process of the biological object" [36]. Therefore, empirically was selected range of optimal frequency electrical impulses. Impulse generators have been put in high voltage (the amplitude). For example, if it based on the electrical induction coils or the voltage multiplier [80, 82].

In accordance, with the development of the impulse generator and new concepts of the physical process changed the construction the "ionizer". This essentially became the contour electrode "antenna-emitter" given the frequency of impulses and impulse duty cycle. Consequently, the biological object drying (the condensed skimmed milk) can be considering the "antenna-receiver" of these electrical impulses.

This was confirming by studies in the laboratory conditions using the oscilloscope-based IBM PC computer modeled using microchips under our experimental conditions.

In all experiments, the research applied the high voltage impulses generator, so the question was, how electronic devices react to the electromagnetic radiation of the high voltage impulses generator at all, and its impact in the environmental field, in particular. Therefore, considerable attention in studies on the induction of specific signals (electrical impulses) in the environment, based on the analysis of radio frequency distortion (interference) arising from these signals.

As any the high voltage electrical equipment used in production conditions, produces broadcast signals impulses which deform industrial radio frequencies, then analyzed what remedies are possible from electromagnetic radiation.

2.5. THE INDUCTION OF ELECTRICAL IMPULSES IN BIOLOGICAL OBJECTS WHEN PHASETRANSITIONS. THE INFLUENCE OF ELECTRICAL IMPULSES ON THE ACCURACY OF THE MEASUREMENTS THE TEMPERATURE OF THE SKIMMED MILK IN DURING FREEZING

Professor, Doctor of the Technical Sciences V.V. Ilyukhin from Moscow State University Applied Biotechnology led the research team, which conducted numerous experiments in the scientist field of phase transitions of the different in its structure and the composition of substances and materials. This team received the patent for the research from the Federal Service of the Russian Federation on the intellectual property and patents [68].

In the research laboratory Moscow State University Applied Biotechnology was established experimental setup, using the virtual oscilloscope based on PC computer with the parameters necessary for the research. With this experimental setup was the interesting process by which was the induction electrical impulses from biological objects (substances) in the phase transition of the first kind (the freezing) [36].

Justified the provisional claim (the patent), which had provided the testimonial the open process that "...this phenomenon is that the smallest particles of biological objects (substances or materials) at the molecular level, can initiate the process of the transition phase, the interaction between the particles occurs due to intermolecular forces the minimum energy level.

These particles of biological substances and materials have the ability to induce of weak electric impulses in the form of the sinusoidal (the harmonic) oscillation, which are characterized by the certain amplitude, frequency and duty cycle" [68].

"In the process of the physical phenomena, phase transitions of biological objects (substances or materials) are of two kinds. The phase transition of the first kind of fast changing physical characteristics such as density of biological objects (substances or materials), the concentration of the components of the substance, in the structure of the mass of the same substance synthesis (the selection or the absorption) of the energy in the form of the heat, etc.

The phase transition of the second kind is the process in which the physical quantity, such as the density of the object (the substance or materials) is constantly changing. Then there is the variable component, the energy in the natural form of the heat is the constant.

For example, the phase transitions are the natural phenomenon, and are constantly in physical processes.

Which include such phenomenon, as the freezing and the defrosting, the melting, the evaporation, the condensation and the crystallization, as well as some structural transitions in solids, for example, the emergence of martensite in the alloy of the iron-carbon [38].

For the scientist researching and studying of the phase transition characteristics of biological substances in the first phase is an analysis of existing thermometric devices, which can fix very precisely parameters of thermal characteristics of biological objects (substances). It is necessary to rule out any abnormalities during many experiments. In the industry and in the laboratory research used liquid, mechanical, electrical, optical, gas and infrared thermometric devices, etc [12, 61 and 62]. In this case, it is necessary to take into account minor deviations in the process of measuring the accuracy temperature of biological objects (substances) in the processes with phase transitions of the first kind or of the second kind.

Thus was created the experimental installation (Fig. 5) using that exact measurements are possible in studies during phase transitions.

In the experimental installation, in the thermo-insulating glass (1), this by its characteristics is the pyro - electrical, flooded with liquid H_2O (4). All this, in turn, placed in the caisson (2).

The temperature inside of the caisson (2), is constant and equal to zero degrees by scale of Celsius. In the fluid H_2O 4, of the caisson (2), installed the electrical Sensor-Probe (3) and the thermometric sensor-thermostatic compact cartridge (5). In the thermos (7) this installation, like in principle to the bulb with the double shell, inside of which the vacuum (by type – Dewar vessel), establishes the thermocouple (6). This thermocouple (6), placed in the liquid mixture with ice (8). This mixture has the constant temperature is intended for the comparative process parameters from the thermometric Sensor-thermostatic compact cartridge (5) in this experimental installation.

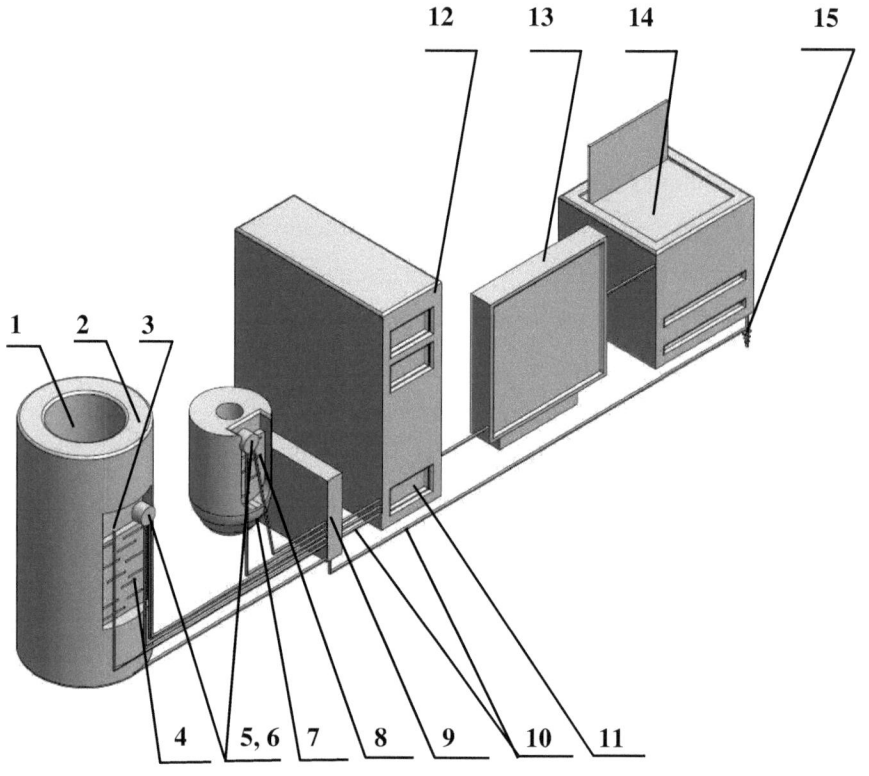

Fig. 5 The experimental installation for computer thermometry
biological objects:

1 – the thermo-insulating glass; 2 – the caisson; 3 – the electric Sensor-Probe; 4 –
the fluid H_2O, 5 – the thermometric sensor - thermostatic compact cartridge; 6 –
the thermocouple; 7 – the thermos; 8 – the liquid mixture with ice; 9 – the adapter;
10 – communication cables; 11 – the chip control; 12 – the processor of the
computer; 13 – the computer monitor; 14 – the print device; 15 – the grounding

Communication cables (10) provide the information of the thermometry by the electrical Sensor-Probe (3) and the thermometric sensor-thermostatic compact cartridge (5), the thermocouple (6), through the adapter (9) to the chip control (11), the processor of the computer (12) [68]. The information can be analyzed on the computer monitor (13) or on paper media by using the print device (14). The adapter (9) of the experimental installation grounded (15). The electric Sensor-Probe (3) is made of copper. The thermometric Sensor-thermostatic compact cartridge (5) and the thermocouple (6) are highly sensitive sensors made of alloy, non-ferrous metals (chromium, copper) and connected with the adapter cables with protective sheath.

Thermometry of liquid H_2O at phase transition has the following algorithm of sequential actions:

1.   There is the process of the freezing (the crystallization) of the liquid in the thermo-isolation glass (1);

2.   On the electric Sensor-Probe (3) at this point, there are electrical signals in the form the impulse of amplitude modulation induced by particles (molecules or groups of molecules) H_2O in the process of the crystallization during phase transition first kind;

3   Electrical impulses are transformed through the adapter (9) to the chip control (11), the processor of the computer (12);

4   After analyzing the impulse amplitude modulation on the computer monitor (13), the information transfer on paper shown in Fig. 6.

Now connect to the thermometric Sensor-thermostatic compact cartridge (5), which already takes sum signal:

1)   By means the thermoelectric effect of Sebeka, in which there is the electromotive force, which occurs due to temperature difference of the solder, made of different metals;

2)   Due to the effect of the phase transition.

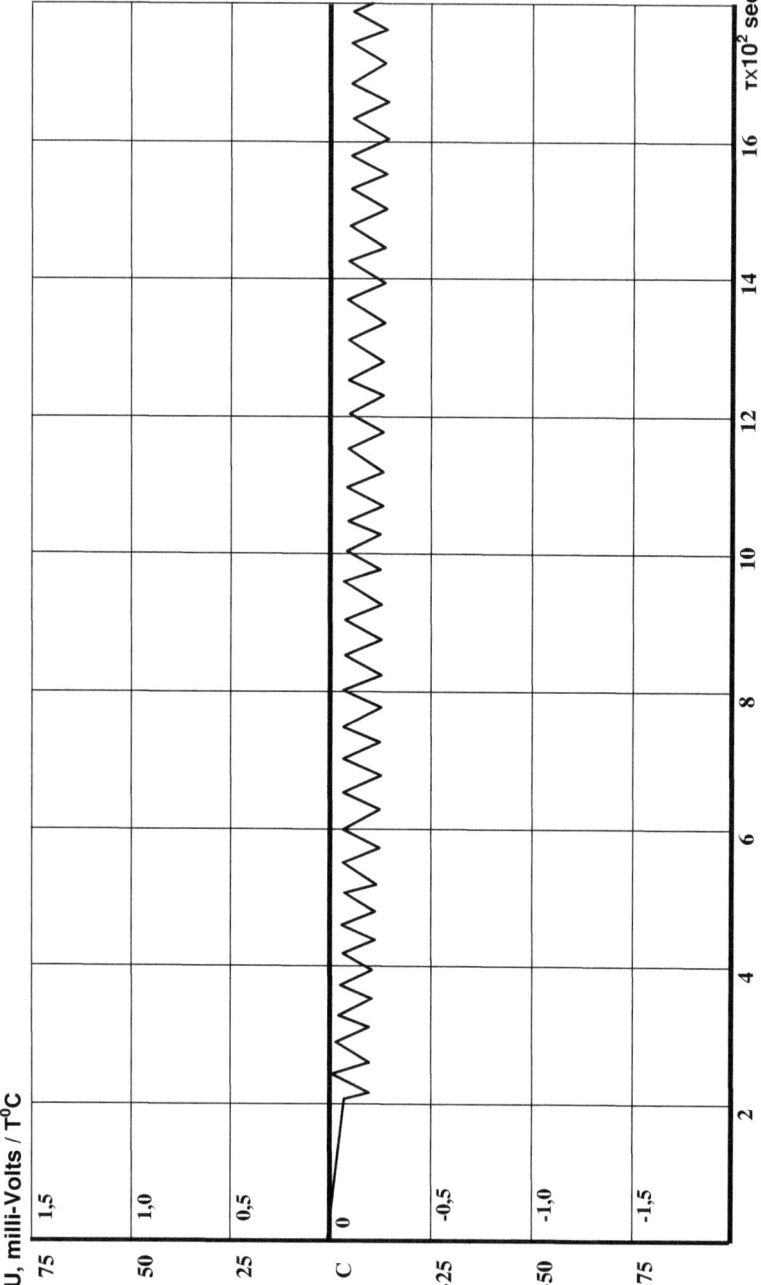

Fig. 6 Diagram the amplitude modulation of impulses in the conversion of liquid to the ice

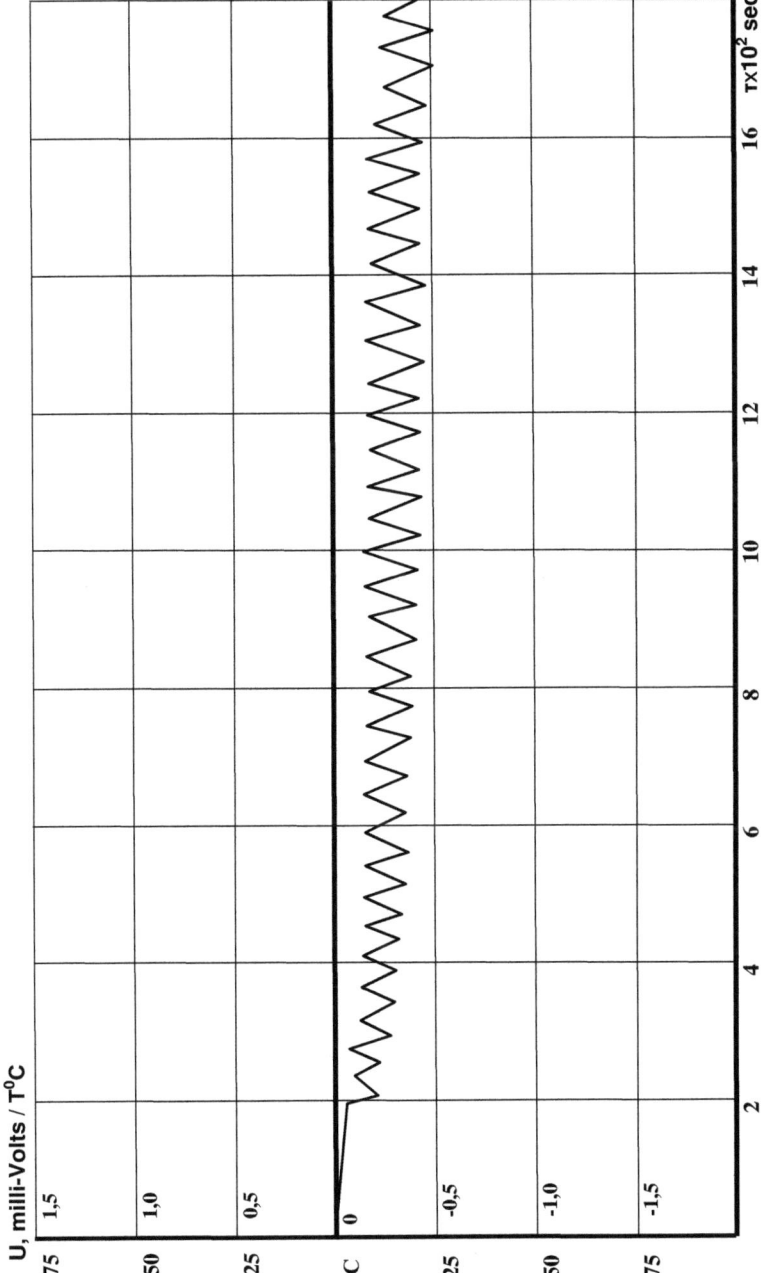

U, milli-Volts / T⁰C

Fig. 7 Diagram the summary amplitude modulation of impulses in the conversion of liquid to the ice

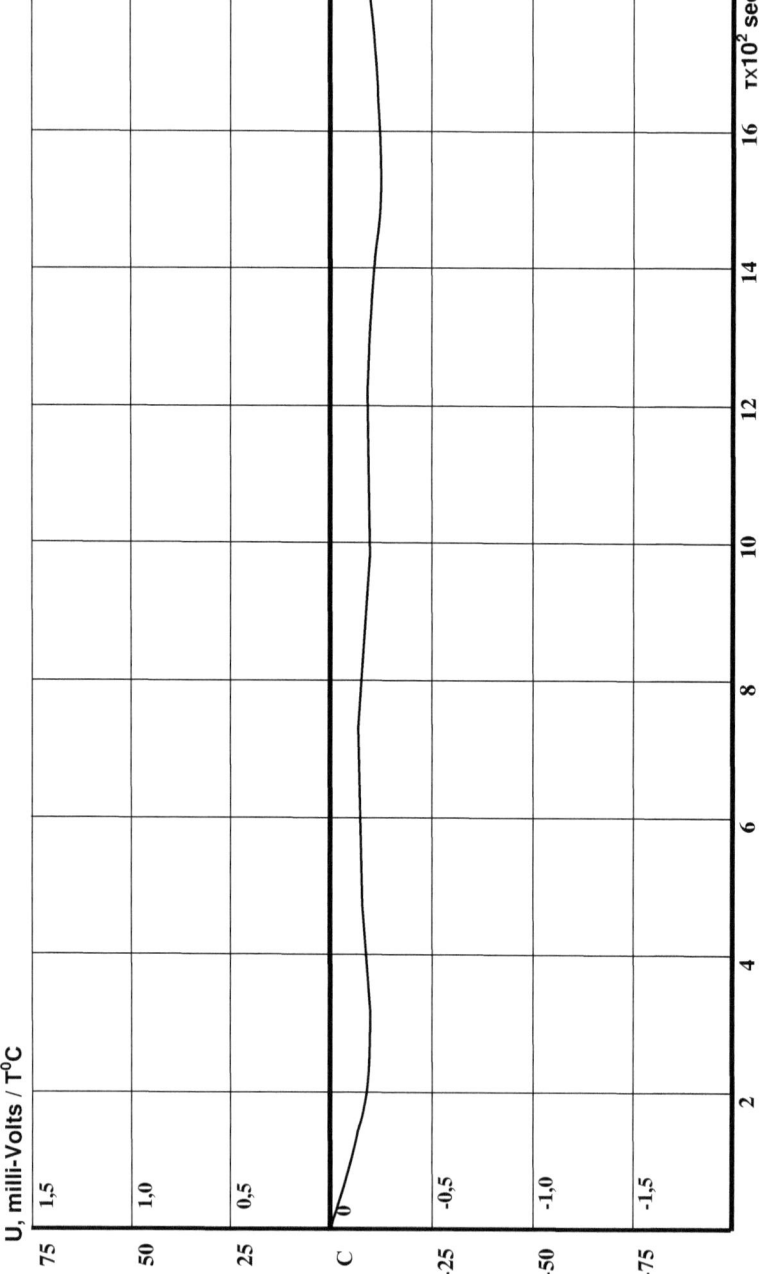

Fig. 8 Diagram disintegration the amplitude modulation of impulses shown in Fig. 6 and Fig. 7.

40

The summary impulse amplitude modulation signal is analyzed on the monitor of the computer (13) and is transferred to the paper through a print device (14). Now with the help of the chip control (11), see to the monitor of the computer (13) and the impulse of the amplitude modulation signal overlay with the phase transition and the summary impulse signal, shown in Fig. 7.

We get curve middle values of the amplitude modulation and fix diagram disintegration of amplitude modulation impulses, shown in Fig. 8.

Diagrams show that this is the correct indicator the temperature of the liquid during the freezing of H_2O, i.e. in during time the phase transition of the first kind.

It may be noted that in studies of heat-mass transfer processes with phase transitions the first kind in industry, like as the evaporation, the drying, the condensation, the freezing, the sublimation, the de-sublimation, the thawing, etc. In this case, the heat, the mass, and the electrical must be considered as the total unit. Otherwise, specify the required admission. All the biological objects (substances or materials) used in the industry need to make correction tables.

That should take into account the parameters of electrical impulses induced by phase transitions. For example, this was done for constants of the heat phase transition of biological substances and materials [17].

This method was tested on thermometry next objects, such as the water, the acetone, the kerosene, mineral substances, and products of the biological origin, for example, the skimmed milk.

Important the following hypothesis, that the biological object is the skimmed milk, is the source of the impulse induction during the phase transition of the first kind, that is, when the moisture evaporates during the drying process. This physical phenomenon could be the basis for the occurrence of the stochastic resonance.

The stochastic resonance is overlap frequency impulses induced by chaotic dynamics from the external source. For example, the impulse frequency of the electrical potential from the external source is "blending" with the impulse frequency of the biological object. This may be one of the important reasons for the intensification of technological processes in the industry, including dairy plants, for example, the convection drying in the electric field.

Since 2004, Russia acceded to the international practice, assessment of the extraneous water in the milk. The State standards (GOST) specifications are written - "indicator temperature during the freezing of the milk" [42, 44 and 72].

This fact makes the dairy plant to comply with the requirements of the standard for acceptance of the row milk and the search for optimal solutions in equipping laboratories necessary measuring devices.

It should be noted that in standard freezing point of the milk (minus 0.520 C^0) is fairly "hard" and the price of failure in its dimension can be quite high.

For determining the extraneous water in the milk using various devices of well-known companies, such as the firm Advantage Instruments, Fiske Ass (United States), CrioStar 1, CrioStar-automatics the firm Funke-Gerber (Germany), the firm FBI (Italy) and others.

However, if Cryo Osprey is designed to measure the temperature of freezing of the milk, and Osmometers, which in principle are suitable to measure the same parameter, mainly used in the biochemical research.

In Russia Federation (before the USSR) there were universal instruments (devices), in the technical documentation which along with the number of parameters of the milk (the fat content, the dry matter, the extraneous water, the protein, etc.), stated the definition of the freezing temperature.

In present time produce out several models Cryo Osprey and Osmometers type MT - 5, MT – 5 - 1, MT – 5 – C with the requirements of the "State standard ISO 30562 – 1997" [50]. However, be aware that devices that detect the freezing point of milk are an indirect method. Parameter or not normalized or not defined the true characteristics of measurement or measurement accuracy is significantly worse than the State standards.

In the research laboratory Moscow State University Applied Biotechnology were researches on the subject of accurate way to measure the freezing point of the skimmed milk by use the experimental installation, which show on Fig. 5.

In the thermo-isolation glass (1) pour of the skimmed milk and the thermo-isolation glass (1) placed in the caisson (2). Thermometry of the skimmed milk at the phase transition carried out in the same way as the water H_2O and has the following algorithm sustained scientific action in research work:

1. There is the process of the freezing (the crystallization) of the skimmed milk in the thermo-isolation glass (1);

2. On the electric Sensor-Probe (3) at this point, there are electrical signals in the form of the impulse of amplitude modulation induced by particles (molecules or groups of molecules) of the skimmed milk in the process of the freezing on phase transition of the first kind [Fig 9];

3 After switch on of the thermometric sensor-thermostatic compact cartridge (5), analyze the summary amplitude impulse modulation (by thermoelectric effect of Electrical Moving Force and the phase transition the first kind) on the monitor of a computer (13) and then print out.

Electrical impulses are transformed through the adapter (9) to the chip control (11), the processor of the computer (12), and then transferred to the print device (14).

Now with the help of chip control (11) this whole process is displayed on the computer monitor (13).

That is, can be seen on the monitor the overlay of the amplitude modulation of the impulse signal during time of the phase transition the first kind and the summary diapason of impulses signals. These diagrams are presented in Fig. 10.

We get curve middle values of amplitude modulation and fix diagram disintegration of amplitude modulation impulses. The curve line of the diagram is presented in Fig. 11.

Diagrams show, that this is the correct indicator of the temperature during freezing of the skimmed milk, i.e. during the time the phase transition of the first kind.

Using this method, temperature measurement accuracy during time the skimmed milk at the phase transition and the traditional measurement method, based on its structural composition can be 5 – 10 milli-Volts, which is up to 30%.

This can be seen based on the analysis of different diagrams, which presented in Fig. 9, 10, 11.

In the process of the drying (moisture evaporation) are electrical impulses with some frequency, duty cycle, and amplitude.

When using the industrial equipment and installations, for the accuracy sensor instruments (devices), which based on the electrical principles of the measuring of the temperature, need to take into account the influence the phase transition of the first kind.

All results must be evaluated with new really positions or need to make corrections in their testimony for the accuracy measurement in industry instruments (devices) [37].

This was for constant heat of phase transition of substances and (materials) and are presented in table adjustment.

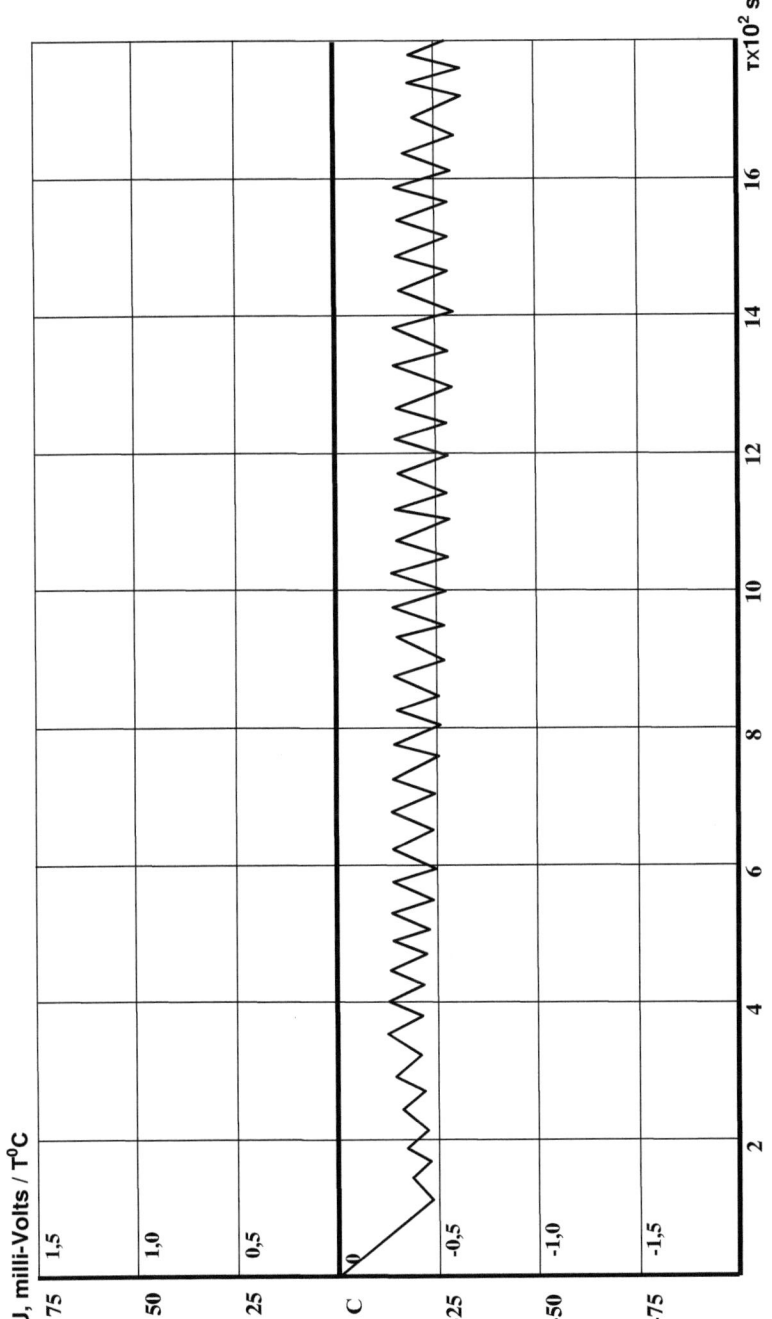

Fig. 9 Diagram the amplitude modulation of impulses for the crystallization (freezing) of the skimmed milk during the phase transition of the first kind

45

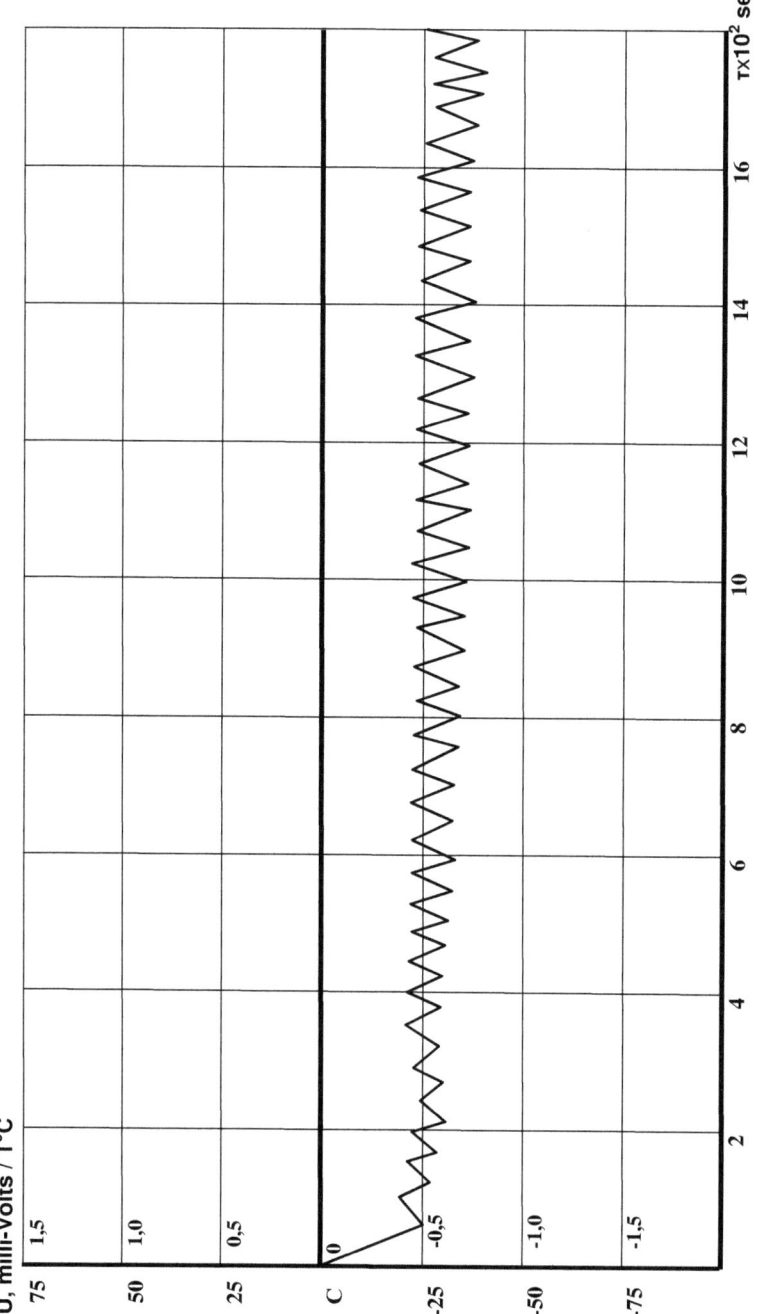

Fig. 10 Summary diagrams the amplitude modulation of impulses for the crystallization (freezing) of the skimmed milk during the phase transition of the first kind and of the thermocouple thermoelectric effects by Electrical Moving Forces

46

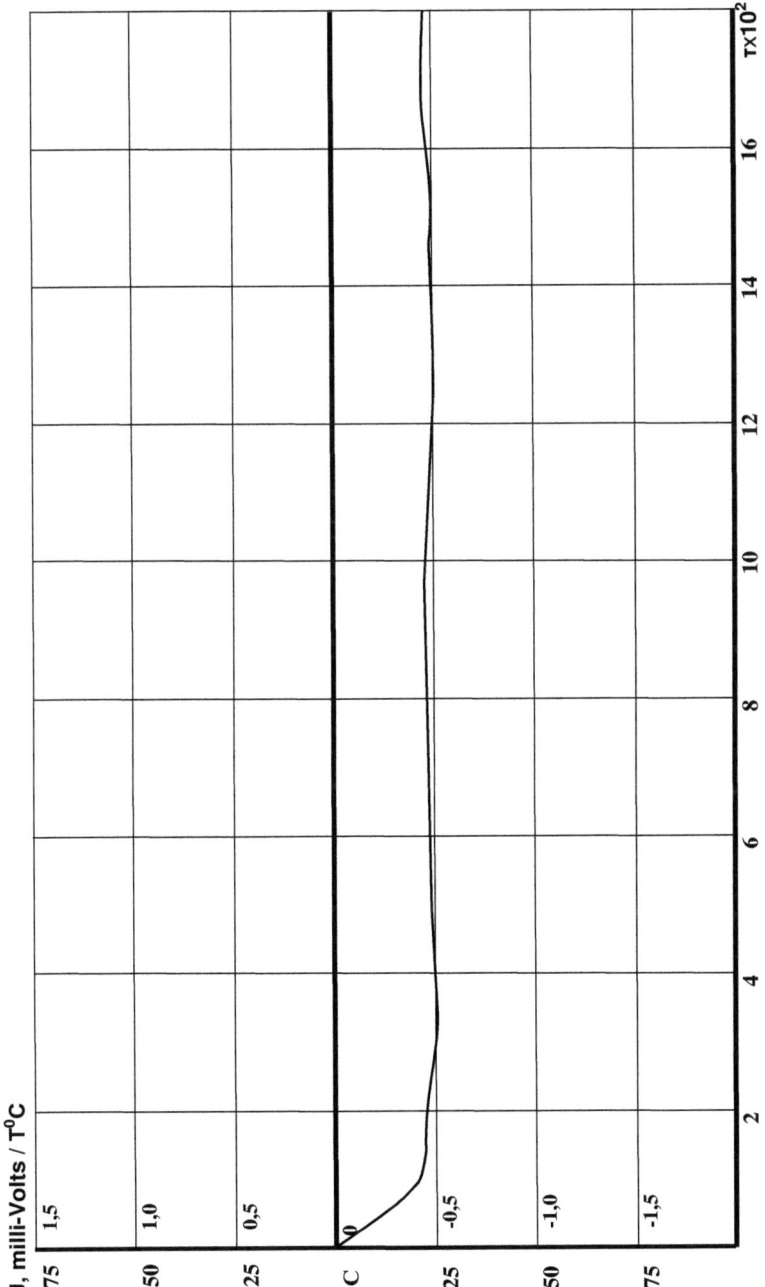

Fig. 11 Diagram disintegration the amplitude modulation of impulses of the skimmed milk shown in Fig. 9 and Fig. 10.

2.6. THE STUDYING AND THE RECEARHING THE AERODINAMICS OF THE ELECTRICAL WIND

Russian and foreign researchers and scientists experimentally determined the effect of electrical fields on an intensification of heat-mass transfer processes [77, 84 and 86]. In the research laboratory Moscow State University Applied Biotechnology also conducted research experiments confirming the effect of intensification of heat-mass transfer process on the drying of biological objects using the influences of weak electrical impulses, the air speed is not changed, and the additional energy to biological objects not using [33, 34].

These studies were based on the hypothesis [11] at the foundation, which was based on the hypothesis that the intensification of heat-mass transfer processes can occur due to the electro-gas convection, as well as by electro-kinetic processes. It is well known that, only after ascertaining the reasons for the increase, you can really manage the process, so the research was carried out on both fronts. In this regard, series of special studies to identify control aerodynamics of the electric wind, one of the main components of heat-mass transfer process using electric potential.

For example, for approximate calculation of the flow speed of gases in the "corona" discharge (the flow of gases into the technical literature is called "electrical wind") provided voltage measurement using the (kilo-Volt-meter). It measures the distance between the electrodes, and then calculating the speed of electrical wind.

Under the direct measurement of the pressure measurement directly understood by means of the graduated tube with fluid, and indirectly, when the necessary amount of the pressure is determined by other options, such as electrical.

2.6.1. THE DEVELOPMENT OF THE INSTALLATION
FOR MEASURING OF THE PRESSURE AND
THE SPEED THE JET FLOW OF THE
ELECTRICAL WIND

We developed the installation (Fig. 12) for measuring the pressure and the speed of the electrical wind, is related to the electric measuring instrument of the aerodynamic testing, which can be used in dryers and heat-exchange devices that use the influence of weak electrical impulses. For the development of this installation have been analyzed already known devices (instruments).

The drawback of this installation is that the speed of the electrical wind measurement is not the direct and the indirect method. That is, a number of assumptions that do not allow you to quickly measure the value of pressure of the gas flow and the speed of electrical wind. It occurs in conditions of the "dark" discharge, without the formation of the plasma or independent discharge which is accompanied by the "corona", which formed by the ionized gas.

There is the installation, which performs the measurement by lowering the temperature of the sensor's thermal resistance, which is flushed with circulating air, according to the scale [18, 74]. Also, analyze the known device for measuring air speed TTM-2-01M. Graduation termo-anemometres TTM-2-01M carried out by the manufacturer without taking into account the effect of the electrical field, so using it in an electric field, inevitably leads to inaccurate results.

Therefore, the appliance termo-anemometres TTM-2-01M the speed electrical wind cannot be used in the electrical field, in view of the fact that the sensor (probe) the reacts very significantly on the electrical discharge in the electrical field.

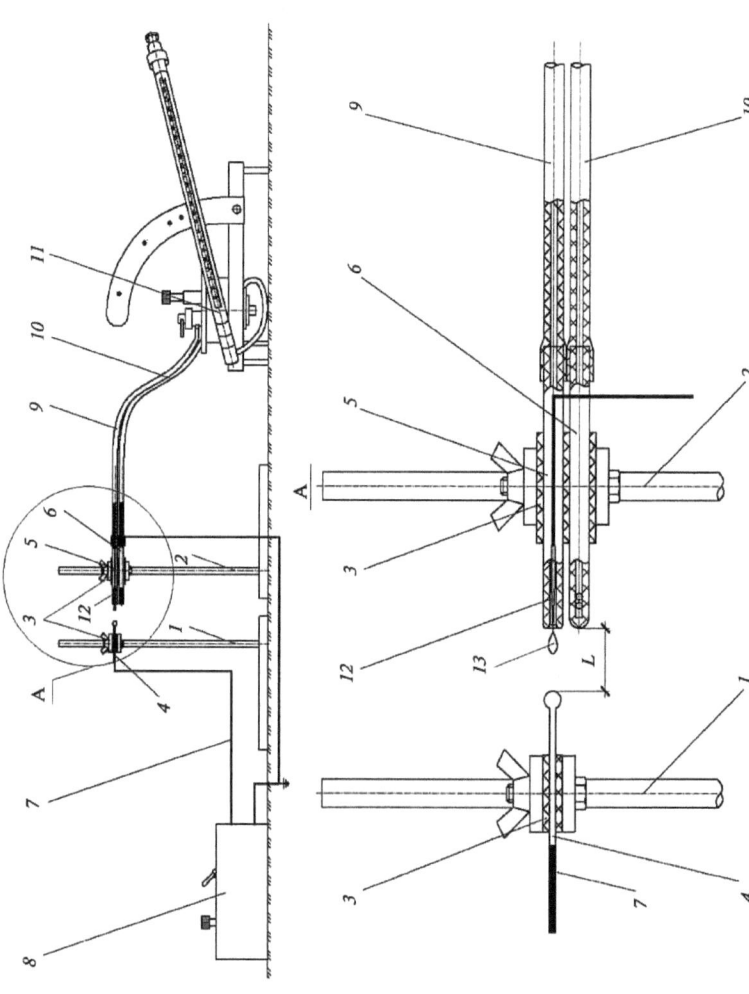

Fig. 12 Installation for measuring the pressure and the speed of the electrical wind:
1, 2 - the tripod; 3 – the electrical insulator; 4 – the spherical electrode; 5, 6 – Pitot tubes; 7 - wire; 8 – the high-voltage electrical current source; 9, 10 - rubber hoses; 11 – micro-manometer; 12 – needle electrode; 13 - zone electric discharge

50

Studies on the possibility of using the installation device is by name "cup-anemometer MC-13". The measurement of the speed electrical wind has shown, that this installation does not react to perturbations (changing) the speed and pressure generated by the electrical wind.

Therefore, when you develop the new installation to measure the pressure of the gas flow and the speed of the electrical wind was electrical analog installation for measurement of the pressure and flow rate of the gas, including Pitot tubes [41, 43] and the micro pressure gauge model MMN-240, which enables direct measurement of extremely small pressure parameters of the electrical wind.

The drawback of this installation is that without engineering original changes and cannot be used for measuring the pressure of the electrical wind and speed of the electrical wind. The purpose of the scientific research was to develop the installation, which allows direct method to measure the pressure of the gas flow or the electrical wind, and knowing you can determine or to calculate the empirical formula of the speed electrical wind.

The task was solved the scientific collective with the help of the installation comprising Pitot tubes, one of which takes the total pressure and another tube static pressure, connected to micro pressure gauge. In micro pressure gauge for measurement of the gas flow in the electrical field of tubes are made of the dielectric, and the ends are made from the metal.

It is connecting with the necessity to change the polarity of the electrical current supply high voltage electrode. The installation is additionally equipped of two electrodes. The needle electrode fixed in alignment with central hole of the tube, which recognizes the total pressure. This may change the distance from the end of the tube until the tip of the needle. The special wire to the source of the electrical current connects end of this electrode and the other electrode connected to the metal ends of the Pitot tubes.

Experimentally determined, that the radius of curvature of the surface of the butt of the tube should be perform in 10 times greater than the radius of curvature of the needlepoint electrode. The ability to change the distance from the butt of the tube until of the needlepoint connected with the necessity to change the intensity of the electrical field, which is determined according to the formula (1).

$$E=\Delta U/L, \tag{1}$$

Where: ΔU – voltage difference, Volt;

L – length (distance), m.

The micro pressure gauge model MMN-240 is based on the fact that it enables direct measurement of extremely small electrical wind pressure settings using Pitot tubes, rather than indirectly, as in the device (instrument) TTM–2–01M.

The installation for measuring the pressure and the speed of the electrical wind (Fig. 12) made in the next construction. On the tripod (1) mounted on the electrical insulator (3) of the spherical electrode (4), which connected of the wire (7) to the high-voltage electric current source (8). On the tripod (2) this installation on the electrical insulator (3), mounted Pitot tubes (5) and (6).

The Pitot tube (5) considers the full pressure mounted of the needle electrode (12), axle with spherical electrode (4) and through the rubber hose (9) is connect with the micro pressure gauge model MMN-240. To the micro pressure gauge model MMN-240 through the rubber hose (10) connected with Pitot tube (6), which perceiving the static pressure. It is the base of Installation for measuring the pressure and the speed of the electrical wind.

When you turn on high-voltage electrical current source (8) formed the "corona" discharge the needlepoint electrode (12) in the form the Jet of plasma of certain length in zone of the electrical discharge (13) of the electrical field.

From this arises the so-called the "electrical wind", which penetrates into the tube Pitot (5) and then through by the rubber hose (9) to the micro pressure gauge model MMN-240.

This micro pressure gauge detects pressure of the electrical wind readings on the scale of the device. Then carry out the calculation of values the speed of the electrical wind.

Changing the certain length (the distance) (L) and the amount of the supplied voltage of the high-voltage electric current source (8) on the spherical electrode needle (4) to determine the pressure of the electrical wind by means scale of the micro pressure gauge model MMN-240.

As it was already said that, under the direct measurement of the pressure measurement directly understood by means of the graduated tube with fluid, and indirectly, when the necessary amount of the pressure is determined by other options, such as electrical

The speed of the electrical wind shall be calculated by the formula **2**:

$$\omega = \sqrt{2gH}/\sqrt{\gamma} \qquad (2)$$

Where: ω –   speed of the electrical wind, m/sec;

g –   acceleration of gravity, m/sec²;

H –   dynamic pressure, Pascal;

γ –   specific weight of the air, which depends from the voltage of the electric field, kg.

2.6.2. THE DEPENDENCE OF THE PRESSURE AND THE SPEED OF THE ELECTRICAL WIND FROM THE VOLTAGE OF THE ELECTRICAL FIELD

Developed the installation for measuring the pressure and the speed of the electrical wind can be use in heat-exchange apparatus and the equipment, not only in the food industry, but also in other industries, for example, in the refrigeration technology using the influence of weak electrical impulses.

Laboratory tests defined based on the pressure and the speed of the gas flow (the electrical wind) from the voltage of the electrical field at independence and dependence discharge.

Under independence discharge, understand electrical "corona" discharge accompanied by glowing plasma formed in the form of the ionized gas. By dependence discharge (dark) understand category with less the voltage the electrical field and is not accompanied by the formation of the "corona" (plasma).

In the Table 1. fixed laboratory scientist studies on the measurement of the pressure and the speed of the electrical wind with depending on the intensity from the electrical field.

The diagram pressure of the flow electrical wind from the voltage of the electrical field without extreme transitions is presented in Fig. 13.

When the dependence discharge, pressure changes from 1,55 to 3,87 Pascal, if you change the voltage of the electrical field from 625 to 666,7 kV/m.

When the independence discharge, pressure changes from 3,87 to 11,62 Pascal, if you change the voltage of the electrical field from 666,7 to 869,6 kV/m. accordingly.

The diagram is presented in Fig. 14 of the speed the flow electrical wind from the voltage of the electrical field shows, that in the transition from dependent to independent electrical discharges has the place the extremum near 666.7 kV/m.

Table 1

The influence of the electrical field intensity on the pressure and the speed of the electrical wind

Variable parameter	The numerical values of the pressure and the speed electrical wind rate depending on the intensity of the electrical field			
Voltage of the electrical field, E kV/m	625	666,7	750	869,6
Pressure of electrical wind or flow of gas, Pascal	1,55	3,87	7,72	11,62
Speed of the electrical wind, m/sec	0,08	0,13	0,17	0,22

When the independence electrical discharge, pressure changes from 3,87 to 11,62 Pascal.

Need to change the voltage of the electrical field from 666,7 to 869,6 kV/m. accordingly

The gas flow the speed (the electrical wind) ranges from 0.08 to 0.13 m/sec when the voltage of the electrical field from 625 to 666.7 kV/m. The maximum distance of the electric wind in voltage of the electrical field 869.6 kV/m and 20 kV voltage is only 0.023 m.

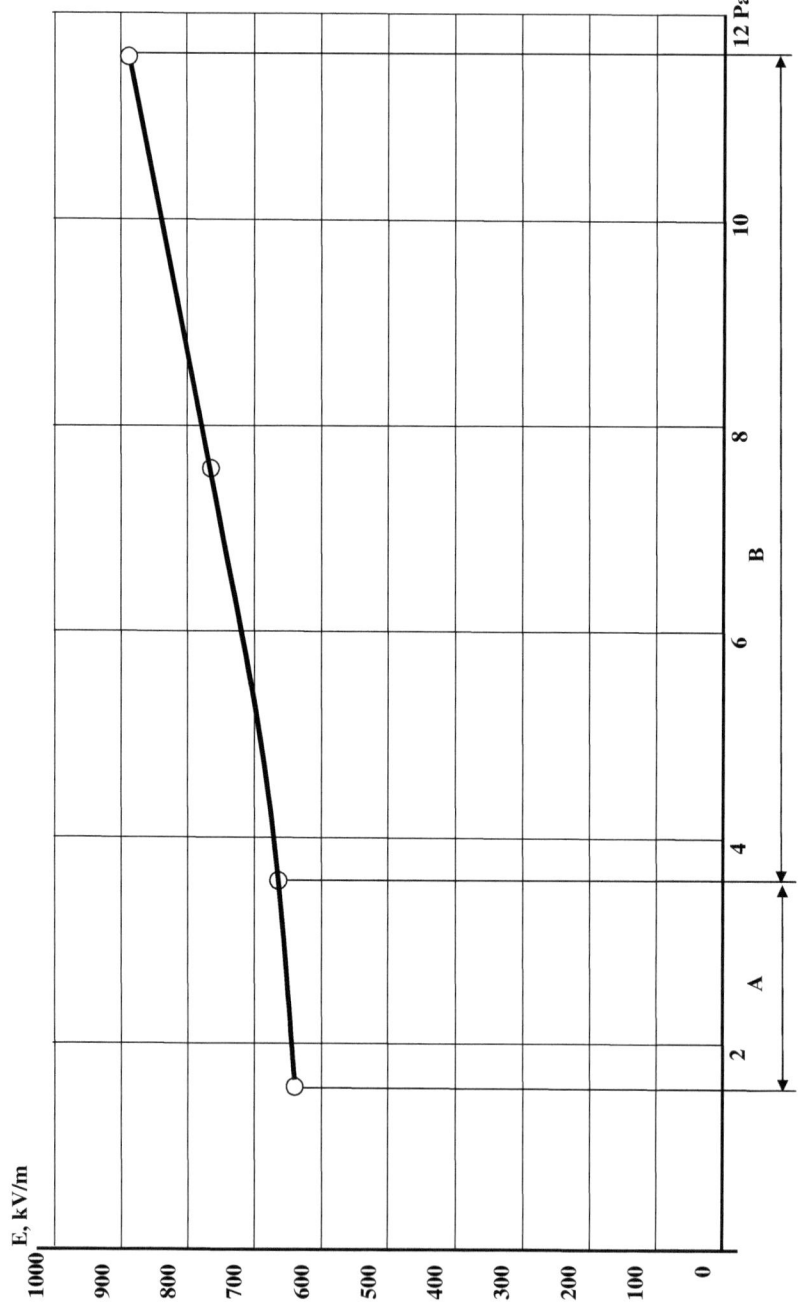

Fig. 13. Diagram change pressure of the flow electrical wind
A - zone dependence discharge; B - zone independent discharge

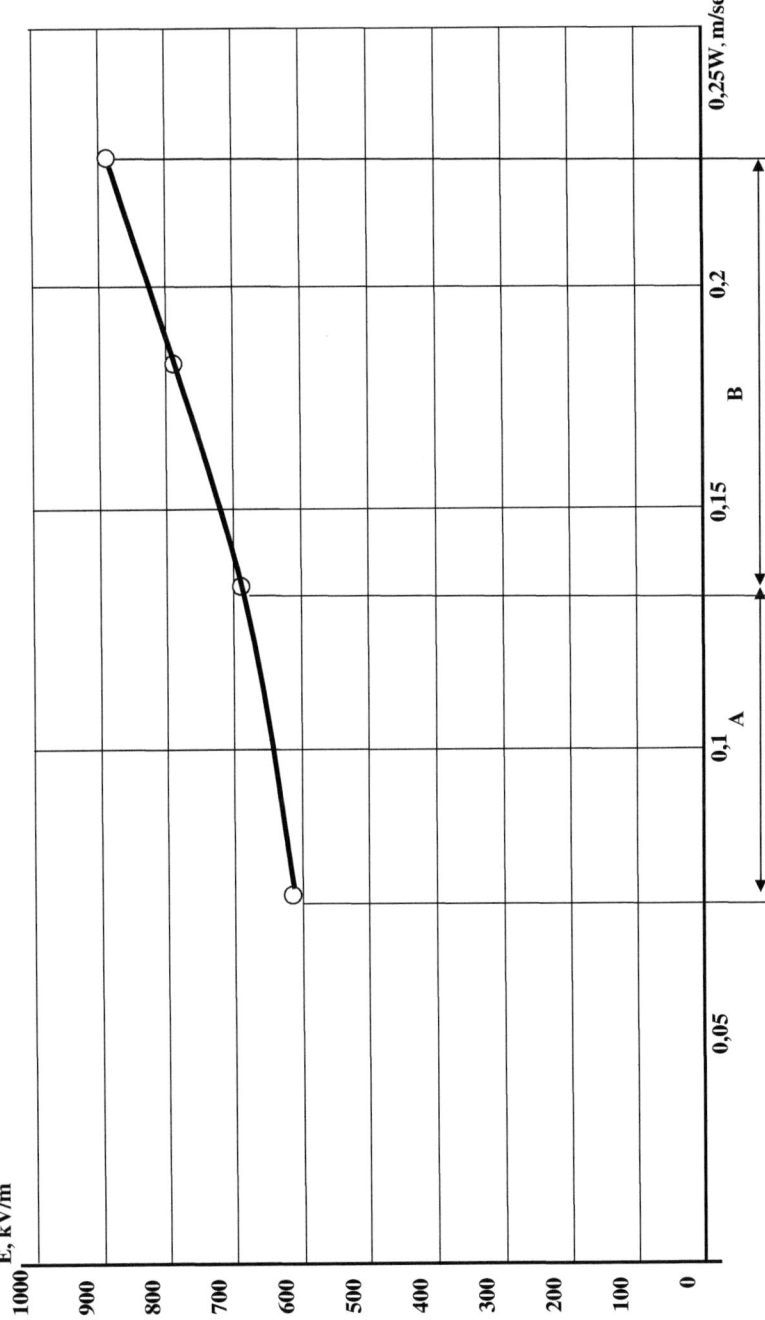

Fig. 14. Diagram change the speed of the flow electrical wind
A - zone dependence discharge; B - zone independence discharge

Fig. 16 shows the scheme of the Jet flow and the suction torch electrical discharge:

Where: U   – speed of the flow of the electrical wind, m/sec,

U_0 – beginning speed the flow of the electrical wind, m/sec,

U_m – axial speed the flow of the electrical wind, m/sec.

On the scheme the outlines of the Jet flow showing in as straight lines. In really, tangent lines, which to indicate voltage, show that existing along the outlines of the Jet flow with a stationary ambient air is generating the vortex. These vortices provide transfer of momentum and impurities (and warmth for non-isothermal jets) from the Jet flow in ambient air and back.

Experimental studies have shown that the component of the speed electrical wind on the axis "y" is many times less than on the axis "x", therefore, we can assume that the vector of the speed at any point of the Jet flow is directed along the axis "x".

In order to calculate the movement of gas molecules in the air stream directly beside openings, it is necessary to make certain exceptions.

This does not prevent the version of the experimental researching. The research of the pressure and the speed of the electric wind, the results showed that the efficiency of the "gas electric convection" is quite low.

In addition, temperature measurement carried out with the help of thermocouples, which would inevitably lead to mistakes induced by the electrical induction [37].

The installation for measuring the pressure and the speed of the electrical wind issued patents from the Federal Service of the Russian Federation on the intellectual property and patents [39, 40, 70 and 71].

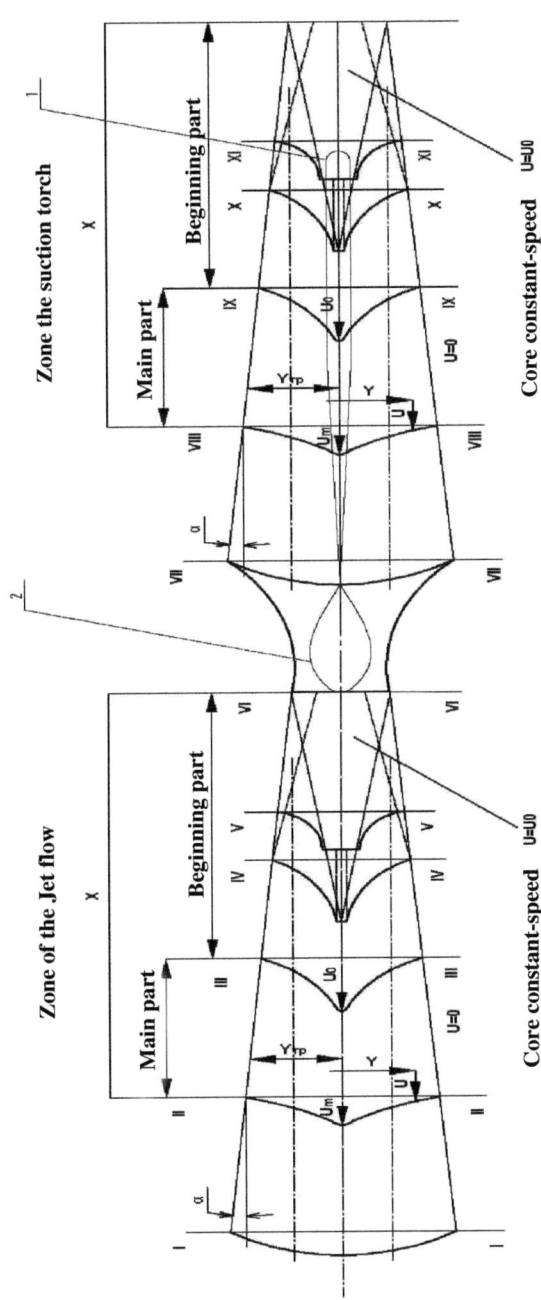

Рис. 15. Scheme of the Jet flow and suction torch electrical discharge:
1 – the needle electrode; 2 – zone of electrical discharge ("corona" when independence discharge or dark discharge)
I – I, II – II, III – III, IV – IV, V – V, VI – VI, VII – VII, VIII – VIII, IX – IX, X – X, XI – XI. – the scheme of speeds

59

CONCLUSION

In this monograph considered the "corona", which was the basis for the further study of the mechanism the intensification of the process drying of biological objects. It was analysis the influence of the "corona" discharge on the state of the ambient air, in addition, shown the positive and the negative influence on the process of the drying. Experimentally selected contour electrode "antenna-emitter", the name of which in other technical sources are classified, as "Ionizer".

The most important stage for further research was developed by the high-voltage impulse generator with optimal characteristics. These devices begin to produce in series for various industries in Russia and used in the drying process.

Studied the high-voltage sources in the form of electromagnetic waves (radiation), which is the high-voltage impulse generator on other sources (devices), operated in the range of these waves, as well as on the environment and the staff. There are recommendations for the protection of these radiations. There are recommendations for the protection of these radiations.

Experimentally determined the pressure and the speed of electrical wind is as well as a method of determining the pressure Jet flow of gas also through experimentation.

Formulated the hypothesis, which can to explain the mechanism the influence of the electric field intensity are due to the convective drying of electro - kinetic processes, the "stochastic resonance" and "white noise".

BIBLIOGRAPHY

1   Anderson O. Physical acoustics. – Moscow: Publishing House "Peace". – 1968. – 391 p.

2   Andrianov E.I. Definition methods structurally mechanical characteristics of powdery materials. – Moscow: Chemistry, 1982. – P. 121 – 152.

3   Anisimova V.I. Research of x-ray radiation at destruction of adhesive contact / V.I. Anisimova, B.V. Deraygin, V.A. Kinoev // Materials V of the All-Union symposium on mechanic-issue and mechanic-chemistry of solid bodies. – Tallinn, Estonia: 1977. – P. 98 – 103.

4   Babayev I.E. Research of process and development of the equipment of continuous sublimation drying of the granulated mincemeat in a vibration layer: author's abstr.... Cand. Tech. of Sciences. – Moscow: 1976. – 31 p.

5   Basov A.M. The main directions the of electron-ion technology of processes of the agricultural production // Theses of reports of the II All-Union conference on application of electron-ion technology in the national economy. – Tbilisi: 1978. – P. 87 – 88.

6   Bazelayn E.M. Spark of the discharge. / E.M. Bazelayn, Yu.P. Raiser – MFTI, Moscow: 1997. – 319 p.

7   Beglaryan R.A. Improvement the technology of dairy products influence of magnetic field on milk and lactic acid bacteria: author's abstr.... Dr. Tech. of Sciences. – Moscow: 1989. – 49 p.

8   Bessonov L.A. Theoretical bases of electrical equipment. – Moscow: Higher School, 1967. – 775 p.

9   Bezzubov A.D. Ultrasound and its application in food industry. / A.D. Bezzubov, E.I. Garlinskaya, V.M. Friedman. – Moscow: Food Industry. – 1989. – P. 112 – 118.

10   Boot A.I. Electron-ion processes of water structures of live organisms and products of their processing. – Moscow: 1992. – P 110 - 112.

11   Bredov M.M. Electrization, found after contact of two solids. / M.M. Bredov, A.E. Ksheminskaya // Journal of Technical Physics. – Moscow. 1957. –T. 27. –P. 921 – 925.

12   Brout R. Phase transition // University of Brussels, New York – Amsterdam. 1965. – P. 7 – 9.

13   Brout, R. Phase transition // University of Brussels, New York – Amsterdam. 1965. – P. 7 – 9.

14   Bulkin M.S. Sublimation drying of raw materials of a biological origin taking into account fluctuations in industrial technologies author's abstr.... Dr. Tech. of Sciences. – Moscow: 2010. – 54 p.

15   Burlev M.Ya. Ecological aspects of use of the "corona" discharge for processing of agricultural products. / M.Ya. Burlev, V.V. Ilyukhin, S.S. Ilyukhina, N.V. Makarov, I.M. Tambovtsev, S.V. Shishkin // Increase of efficiency in processing industries of agrarian and industrial complex: Scientific works. "75 anniversary of Moscow State University Applied Biotechnology" "It is devoted to century since the birth of Professor A.N. Lepilkin". – Moscow: State University Applied Biotechnology, 2004. – P. 50 – 55.

16   Burlev M.Ya. Production of the powdered skimmed milk with using of weak electric impulse influences: author's abstr...Cand. Tech. Science. – Moscow; 2002. – P. 15 – 17.

17   Burlev M.Ya. Intensivierung des Prozesses von dehydratisieren in elektrischem Feld schwaches Impulse / M.Ya. Burlev, N.S. Nikolaev // Science and Education. Materials of the III international research and practice conference. Vol. 1. Publishing office Vela Verlag Waldkraiburg. – Munich, West Germany. April 25th – 26th, 2013. – S. 88 – 91.

18   Cady, W. Piezoelectricity an introduction to the theory and applications of the electromechanical phenomena in crystals. New York – London. 1946. – 716 p.

19   Carlin B. Ultrasonic. / Technical application // New York – Toronto – London: 1949. – 305 p.

20   Cavendish – Priestly. Patent / Scientific magazine // History and Present State of electricity, with Original Experiments. USA, Great Britain: 1784. – 71 p.

21   Cavendish – Priestly. Patent / Scientific magazine // History and Present State of electricity, with Original Experiments. USA. Great Britain. 1784. – P. 71.

22   Chizhevsky A.L. Aero ions and life. – Moscow: Publish House, 1989. – P. 322 – 344.

23   Colodeznaya V.S. Use of ozone at refrigerating storage of products of the animal origin: author's abstr...Cand. Tech. of Sciences. – Leningrad, Moscow: 1975. – 22 p.

24   Cooke G.A. Processes and devices of the dairy industry. – Moscow: Food Industry. – 1973. – 767 p.

25   French patent Lefebvre No 31470. 1903. – 12 p.

26   Green, H.S. Order – Disorder Phenomena / Green H.S. Hurst C.A. London. Great Britain. 1964. – 178 p.

27   Ginsburg A.S. Basic theory and technology of drying of food products. – Moscow: Food Industry. – 1973. – 664 p.

28   Ginsburg A.S. Drying foods: training manual for engineering in the food industry. – Moscow: Food Publish. – 1960. – 683 p.

29   Glushchenko N.A. Bases of the theory and the practical of electro aeration of solutions in food biotechnology: author's abstr. ... Dr. Tech. Science. – Moscow: 1988. – 44 p.

30   Grigoriev V.V. The main directions of electron-ion technology processes of agricultural production // Theses of reports of the II All-Union conference on application of electron-ion technology in the national economy. – Tbilisi: 1978. – P. 2 – 4.

31   Guygo E.I. Sublimation drying in the food industry / E.I. Guygo, N.K. Zhuravskaya, E.I. Kaukhcheshvili. – Moscow: Journal "Food Industry". – 1972. – 425 p.

32   Ilyukhin V.V. Physic technical basis Cryo food cutting. – Moscow: "Agro – Publish". – 1990. – 250 p.

33   Ilyukhin V.V. Drying with use of the electron-ion technology. / V.V. Ilyukhin, M.Ya. Burlev // Journal "Dairy Industry". – Moscow: 1992. – No. 3. – P. 41 – 44.

34   Ilyukhin V.V. Measurement of the temperature of the milk - raw material by Cryo Osprey / V.V. Ilyukhin, M.Ya. Burlev, I.M. Tambovcev, S.V. Shishkin, S.S. Ilyukhina // Moscow, Journal "Dairy industry". – 2005. – No. 12. – P. 40 – 41.

35   Ilyukhin V.V. Phenomenon generating electric impulses in phase transitions. / V.V. Ilyukhin, M.Ya. Burlev, I.M. Tambovcev, S.V. Shishkin, S.S. Ilyukhina / Bulletin of the International Academy of Refrigeration. – Sankt Petersburg. – 2005. – № 4. – P. 15 – 17.

36   Ilyukhin V.V. Production of powdered skimmed milk with use of weak electric impulse influences. / V.V. Ilyukhin, M.Ya. Burlev. – Moscow: Journal "Dairy Industry". – 2001. – No. 9. – P. 57 – 58.

37   Ilyukhin V.V. Stochastic resonance unstable brought electrical impulses in the processes of cooling, freezing and drying of biological objects. / V.V. Ilyukhin, M.Ya. Burlev, S.V. Shishkin S.S. Ilyukhina // "Food, ecology, people: proceedings of the 5-th international scientific and technical conference". – Moscow, 2003. – P. 215 – 216.

38   Ilyukhin V.V. Using in technology of produce of bio-composite materials phenomena synchronization of unipolar electric impulses substances, phase transitions first kind. / V.V. Ilyukhin, M.V. Lekishvili, M.Ya. Burlev, M.B. Zyankin // Actual problems cell transplantation: Collection of abstracts of IV all-Russian Symposium. – Sankt Petersburg, 2010. – P. 68 – 70.

39   Ilyukhin, V.V. Aerodynamics pressure of the Jet flow and of the suction torch of the electrical discharge. / V.V. Ilyukhin, M.Ya. Burlev, E.V. Zhukavets // Bulletin of the International Academy of Refrigeration. – Sankt Petersburg, 2012. – № 3. – P. 20 – 22.

40   Ilyukhin, V.V. Aerodynamics pressure of the Jet flow electrical wind and of the suction torch of the electrical discharge. /V.V. Ilyukhin, M.Ya. Burlev, E.V. Zhukavets // Journal "Dairy Industry". – Moscow, 2011. – № 4. – P. 83 – 84.

41   ISO 15528-86. The means of measurement, volume or mass flow rate of liquid and gas.

42   ISO 52054-2003. "Natural cow's milk – raw of milk".

43   ISO 5764-1997 "Milk. Determination of the freezing point. Cryo Osprey method by Thermistor".

44   ISO 8361-79. Digital Anemometer. Fluid and gas consumption. Methods of measurement for the speed at one point section of the pipe.

45 Ivashov V.I. Technological equipment of enterprises of the meat industry: – Part 2. Equipment for processing of the meat. – Sankt. Petersburg: GIORD. – 2007. – 106 p.

46 Ivashov V.I. The laboratory practical work: manual for students. / V.I. Ivashov, S.G. Yurkov, V.V. Ilyukhin, B.N. Duydenko, V.A. Katyukhin. – Moscow: MTIMMP, 1987. – P. 60 – 75.

47 Kachurin L.G. Laboratory works on experimental physics of the atmosphere / L.G. Kachurin, A.N. Merzhevsky. – Leningrad: Gidro-Meteo Publishing, 1969. – 500 p.

48 Kaptsov N.A. The electric phenomena in gases and vacuum. – Moscow-Leningrad: Publishing House of Technical Literatures, 1950. – 825 p.

49 Kharitonov V.D. Scientific basis and practice the technology of powdered milk by method of two-phase drying: author's abstr. ... Dr. Tech. Science. – Moscow: 1989. – 46 p.

50 Kirsanov V.I. Temperature measurement the freezing of the raw milk. // Moscow, Journal "Dairy Industry". – 2004. – № 9. - P. 18 – 19.

51 Krapivina S.A. Plasma - chemical technological processes. – Leningrad: 1981. – 245 p.

52 Lang O. / Vermeidung von Gewichtsverlusten beim Kuhlen und Gefrieren von Fleisch // Die Fleischwirtschaft. Germany: 1978. No. 3. – S. 402 – 405.

53 Lipatov N. N. Powdered milk / N. N. Lipatov, V.D. Kharitonov. – Moscow: Light and Food Industry. – 1981. – P. 3 – 4, 85, 116, 160, 200, 210, 212, 263, 300.

54 Lupu O.F. Theoretical and experimental study of the drying process of apricot with using high-frequency currents: author's abstr.... Dr. Tech. of Sciences. – Chisinau: 2010. – 54 p.

55   Lurie M.Yu. Drying. – Moscow: Energy Publish – 1978. – 711 p.

56   Lykov A.V. Theory of heat conductivity. – Moscow: Higher School. – 1967. – 244 p.

57   Lykov M.V. Spray dryers. / M.V. Lykov, B.I. Leonchik. – Moscow: Mechanical Engineering. – 1966. – 336 p.

58   Masters K. Spray Drying Handbook. – George Godwin. London. Great Britain: 1985. – P. 696 – 701.

59   Novikov Yu.N. Electric technical and electronic engineer. – Moscow: Science, 2005. – 380 p.

60   Novitskaya N.S. Application of ozone technologies when processing milk. // Milk Processing. – Moscow: 2009. – No. 10. – P. 54 – 55.

61   Nuzhdin A.C. Measurement in refrigeration technology: Reference guide. – Moscow. Agro-Prom Publish, 1986. – 61 p.

62   Oleynik B.N. Accurate calorimetric. – Moscow. Publishing House of the State Committee of Standards, Measures and Measuring devices. USSR, 1964. – 17 p.

63   Patent for the invention No. 2367862 (Russian Federation). The device for ultrasonic drying / V.N Khmelev, A.V. Shalunov, R.V Barsukov, S.N Sctyganok, A.N. Lebedev. // Demand No. 200 8118796/06. – Moscow: 20.09.2009.

64   Patent No. 1322516 (USSR). Device for control of parameters of electrization of ionized gas flow. / V.V. Ilyukhin, – Published in the Bulletin of inventions. Moscow. – 07.07.1987. – No. 25.

65   Patent No. 1349227 (Russian Federation). Way of production of latex products. // G.K Berman, E.A Belder, B.C Yershov, V.V. Ilyukhin, B.A Mayzelis, A.A Pepelyaev, Yu.D. Onishchenko, V.P Starun, A.V. Ivanov Published in bulletin "Invent, useful Models" Moscow: – 2001. – No. 35.

66   Patent № 101825 (Russian Federation). The installation for measuring the pressure and the speed of the electrical wind. / V.V. Ilyukhin, M.Ya. Burlev, E.V Zhukavets, A.V. Rascoshny. // Application No. 2010128736. – Moscow. – 27.01.2011.

67   Patent № 101826 (Russian Federation). The installation for measuring the pressure and the speed of the electrical wind. / V.V. Ilyukhin, M.Ya. Burlev, E.V. Zhukavets, A.V. Rascoshny. // Application No. 2010137083. – Moscow. – 27.01.2011.

68   Patent № 1423877 (USSR). A way of drying of materials // Ilyukhin V.V. Bulletins of Inventions edition. – Moscow. – 1988. – № 34.

69   Patent № 1500926 (USSR). The device for control of parameters of electrization of the ionized gas stream // V.V. Ilyukhin, V.V. Kostyushov. Bulletins of Inventions edition. – Moscow: 1989. – № 30.

70   Patent № 230097 (Russian Federation). A way to measure the temperature of the substance during the phase transitionsю / V.V. Ilyukhin, M.Ya. Burlev, I.M. Tambovcev, S.V. Shishkin, S.S. Ilyukhina // Application No. 2005112547. Moscow. – 26.04.3005.

71   Patent USA 0433702 Electrical Transformer of Induction Device. Tesla. 05.08.1890.

72   Radayeva, I.A. Influence of a way of drying on quality of powdered milk / I.A. Radayeva, S.P. Shulkina, Zh.Yu. Petrova // Works VNIMI. – 1973. – T. 32. – Page 6 – 8.

73   Rebinder P.A. Physical-chemical mechanics. – Moscow: Science. 1966. – P. 3 – 16.

74   Remizov A.N. Medical and Biological Physics. – Moscow. High School. 1987. – 637 p.

75   Reuss, F.F. Memories de la Society Imperials des Naturalist's de Moscow. – 1809. – No 2. – 327 p.

76   Revnivtsev V.I. Physical bases of electric separation. – Moscow: Nedra, 1983. – 270 p.

77   Rogov I.A. Physical methods of food processing / I.A. Rogov, A.V. Gorbatov – Moscow, Food Industry, 1974. – 583 p.

78   Romankov P.G. Drying in a suspension. / P.G. Romankov, N.B. Rashkovskaya. – Moscow. – Publishing House "Chemistry". – 1968. – 360 p.

79   Sazhin B.S. Bases the technology of the drying. – Moscow: Chemistry, 1984. – 320 p.

80   Schubert H. Powder Technology. // London. Great Britain. – 1975. – No. 11. – P. 107 – 109.

81   Smirnov B. M. Physic of the weak ionization gas. – Moscow: Science, 1978. P. 102 – 110.

82   Steinberger, E.N. Electrical properties of ice / E.N. Steinberger, S. Rahanim // Journal Applied Meteorology. Volume 10. – 1971. – N 3. – P. 595 – 598.

83   Suponina T.A. Use of ozone at refrigerating storage of potatoes: author's abstr. .... Cand. Technical Sciences. – Leningrad: 1979. – 22 p.

84   Surkov V.D. Processing equipment of the enterprises of the dairy industry / V.D. Surkov, N.N. Lipatov, N.V. Baranovsky. – Moscow: Food Industry. – 1970. – 552 p.

85   Surkov V.D. Technological equipment of the dairy industry / V.D. Surkov, N.N. Lipatov, N.V. Baranovsky. – Moscow, Food Industry, 1970. – 552 p.

86   Taneva, S. Japan Journal of Applied Physics. Japan. Tokyo –1963. – v. 2. – No. 12. Steinberger – P. 798 – 804.

87   Tesla, N. The lost inventions. – Moscow: EKSMO, YAUZA, 2009. – P. 145 – 148.

88   The state and a role of water in biological objects. // Materials of a symposium of Academy of Sciences of the USSR. – Moscow: Science, 1967. – 154 p.

89   Widemann, G., Germany. Berlin. Pegging Anniversary – 1852. – No 87. – P. 321 – 323.

90   Zayas Yu.F. Quality of meat and meat products. – Moscow: Light and Food Industry. 1981. – 480 p.